D1321046

l Guide to Food Additives
n

Essential Guide to Food Additives
4th Edition

Edited by

Mike Saltmarsh
Inglehurst Foods Limited, Alton, UK
Email: mike@inglehurst.co.uk

RSC Publishing

ISBN: 978-1-84973-560-5

A catalogue record for this book is available from the British Library

© The Royal Society of Chemistry 2013

Published by The Royal Society of Chemistry,
Thomas Graham House, Science Park, Milton Road,
Cambridge CB4 0WF, UK

Registered Charity Number 207890

Visit our website at www.rsc.org/books

Printed in the United Kingdom by CPI Group (UK) Ltd, Croydon, CR0 4YY, UK

Preface

The purpose of the Essential Guide to Food Additives is to provide a legal and technical understanding of the use of additives in food. It is intended to be an objective guide in an area that is often characterised by overheated discussion. The first edition was published in 2000 and was sufficiently successful that two further editions were published, in 2003 and 2008, to incorporate the additions and removals of additives and to recognise amendments to legislation.

The publication of the Food Improvement Agents Regulations in the EU in 2011 introduced a new approach to the legislation and necessitated a complete revision of the chapter on European food legislation. At the same time the opportunity has been taken for a complete revision of the text. All chapters have been rewritten and an additional chapter on the development of European food legislation has been included to provide a historical context to assist understanding of food law as it stands in 2012.

The principle of the Guide has always been that the contributions on individual additives have been produced or edited by experts on that class of additive. The first edition was created with the assistance of some 49 individual contributors from the EU and USA. This edition has benefited from updates by a number of these contributors and new contributions from a further 13 experts. I am extremely grateful to these contributors whose efforts keep the guide accurate and up to date as I appreciate that it is very difficult to find the time for additional projects in a busy work schedule. A list of the contributors to this new edition will be found at the back of the book. As ever, there will be those reading this volume who have additional information on particular additives. Comments and contributions are always welcome and with them the next edition will be better.

Essential Guide to Food Additives, 4th Edition
Edited by Mike Saltmarsh
© The Royal Society of Chemistry 2013
Published by the Royal Society of Chemistry, www.rsc.org

Contents

Essential Guide to Food Additives, 4th Edition
Edited by Mike Saltmarsh
© The Royal Society of Chemistry 2013
Published by the Royal Society of Chemistry, www.rsc.org

Contributor List

Thanks to those who have contributed to sections on individual additives in this edition:

Dr.-Ing Thomas Bauch
Harke Pharma GmbH
Xantener Strasse 1
45479 Mülheim an der Ruhr
Germany

Victoria Betteridge
Tate and Lyle plc
Sugar Quay
Lower Thames Road
London EC3R 6DQ
UK

Janine Binder
Sternchemie GmbH & Co. KG
An der Alster 81
D-20099 Hamburg
Germany

Ailbhe Fallon
Fallon Currie Consulting
29 Bressenden Place
London SW1E 5DD
UK

Essential Guide to Food Additives, 4th Edition
Edited by Mike Saltmarsh
© The Royal Society of Chemistry 2013
Published by the Royal Society of Chemistry, www.rsc.org

Barry Foley
D.D. Williamson Ltd,
Little Island Business Park
Co. Cork
Ireland

Dr. R. Hargitt
British Soft Drinks Association
20–22 Bedford Row
London, WC1R 4EB
UK

Alan Imeson
FMC BioPolymer
12 Langley Close
Epsom KT18 6HG
UK

Michel Maton
Alliance Gums and Industries SA
222, Rue Michel Carré
95870 Bezons
France

Dr. Charles Speirs
Campden BRI
Chipping Campden
Gloucestershire GL55 6LD
UK

Krzysztof Szauman-Szumski
Evesa
Pol. Industrial de Campamento
P.O. Box 103
11300 La Linea (Cádiz)
Spain

Valerie Rayner
Sensient Colours
Old Meadow Road
King's Lynn
Norfolk PE 30 4LA
UK

Prof. Arthur Tatham
Cardiff Metropolitan University
Western Avenue
Llandaff
Cardiff CF5 2YB
UK

Daniel Tregear
Air Products plc
Hersham Place
Walton On Thames
Surrey
KT12 4RZ
UK

Professor Bronek Wedzicha
Department of Food Science and Nutrition
University of Leeds
Leeds LS2 9JT
UK

Rachel Wilson
Leatherhead Food Research
Randalls Road
Leatherhead KT22 7RY
UK

CHAPTER 1

Food Additives and Why They Are Used

MIKE SALTMARSH*[a] AND LYNN INSALL[b]

[a] Inglehurst Foods Ltd; [b] Food and Drink Federation,
E-mail: Lynn.insall@fdf.org.uk
*E-mail: mike@inglehurst.co.uk

Consumers in modern industrial countries expect a wide range of foodstuffs to be available throughout the year. Supermarket shelves are stocked with fresh fruit, vegetables, meats and fish from many different countries; prepared salads, breads, preserved meats, spreads, part-prepared dishes ready to cook, cooking sauces, confectionery, drinks and chilled and frozen prepared meals for reheating. All these products have to survive the journey from harvest to store and, once purchased, to have an adequate shelflife for the convenience of the consumer. In many markets, prepared foodstuffs are increasingly required to offer the choice of specific nutritional properties, such as reduced fat, energy content or salt. Food additives are essential to enable the food industry in all its forms to make foods that meet these increasingly challenging demands.

It has become commonplace in some circles to criticise food additives, and in so doing, to lump them all together and imply that they are unnecessary, unnatural and generally not good. This approach ignores their diversity of nature, application and origin. It is often suggested that they are added for no good reason, which is not the case. In fact, the law requires that additives must perform a technological function in the foods in which they are used and may be added only in the minimum quantities necessary to perform that function.

Many of the materials we now call additives are not a recent introduction to the food industry but have been used in foods for hundreds of years. Once man had worked out how to obtain more food than needed for immediate consumption, he

Essential Guide to Food Additives, 4th Edition
Edited by Mike Saltmarsh
© The Royal Society of Chemistry 2013
Published by the Royal Society of Chemistry, www.rsc.org

started to think about how to preserve some of it for times of shortage, such as drought or winter. And when he had established settled communities with a regular food supply, then preparing food to make it more attractive became desirable. There is evidence that Egyptians were using sulfur dioxide to help to preserve wine over 3000 years ago and the Greeks are known to have used a combination of salt and sodium nitrate to preserve meat in the time of Homer. Over hundreds of years, in Europe in particular, international trade increased the range of ingredients available as merchants introduced crops and food ingredients grown by or native to other cultures. Increasingly, cooks and consumers in Europe benefitted from these new ingredients, able for the first time to make cakes, desserts, confectionery, jams and sauces with many of the flavours and textures with which we are familiar today. New preservation techniques such as canning and freezing were added to the tools available to the food producer, and with these and other types of preparation and prepackaging came the requirement for additives such as thickeners, emulsifiers, stabilisers and sweeteners. The challenges posed by lengthening logistics chains, demands for increased convenience, and more recently for reduced fat with no loss of texture, reduced salt with no loss of shelflife, and freeze–thaw stability, have necessitated dedicated research programmes and the development of these more sophisticated additives.

Most countries have a system of approval of food additives, using either the Codex Alimentarius, national or multinational systems. In the EU all additives permitted for use in foods are given a number, prefixed by the letter E. The intention behind the introduction of this system in 1986 was that consumers would find it easier to recognise the number than some of the lengthier chemical names such as polyglycerol polyricinoleate. However, the introduction of the system allowed critics[1] to suggest to consumers that rather than the E number being a mark of approval and safety, it was in fact a cause for suspicion and alarm. Whilst the intensity of the opposition to E numbers has decreased in the EU since then, its legacy persists with "E" numbers being less common on ingredient labels and names being used in preference.

1.1   What are Additives?

The official definition in Europe of an additive, given in Regulation 1333/2008, is "any substance not normally consumed as a food in itself and not normally used as a characteristic ingredient of a food, whether or not it has nutritive value, the intentional addition of which to a food for a technological purpose in the manufacture, processing, preparation, treatment, packaging, transport or storage of such food results, or may reasonably be expected to result, in it or its by-products becoming directly or indirectly a component of such food".[2]

Whilst in most cases it is clear what is an ingredient and what an additive, and the Regulation helpfully provides some examples of what is excluded, even this definition is not entirely precise. For example, salt is an ingredient, vinegar is an ingredient, but if the acid in vinegar, acetic acid, is used alone it must be declared as an additive. Similarly, lemon juice is an ingredient while citric acid, its characterising acid, is an additive.

Food additives have a variety of different roles but in general terms they have two main functions: At the most basic level, they either make food safer by preserving it from bacteria or preventing oxidation or other chemical changes; or they make food look and taste better or feel more pleasing in the mouth.

The use of additives in food preservation is one of the oldest, if not the oldest traditions. Our forebears may not have thought of saltpetre or sodium nitrate, used as a curing agent, or vinegar (acetic acid) used for pickling as additives, but they were essential in maintaining a longer-term food supply of precious perishable foods. Salt, though not an additive by the modern definition, was the other vital ingredient in these preservation techniques.

Alongside their deliberate incorporation into food many additives are also used as processing aids. These are substances that are added either to an ingredient or during the production process, but that do not contribute to the final product. Thus, sulfur dioxide may be added during the peeling and slicing of apples to inhibit browning during the preparation of apple pieces for apple pie. The sulfur dioxide has no role in the final pie, and indeed the majority is driven off during the cooking of the pie. In this instance, the sulfur dioxide is a processing aid as it performs a function during the manufacturing process and is largely absent from the finished product. In the kitchen many of us would use lemon juice to prevent discolouration but this would have a slight effect on flavour. In the food industry, where production is increasingly specialised and expertise focused at specific sites, it is not unusual for the manufacturer of an end product to buy in many of his supplies as part-processed proprietary ingredients. So additives may be needed at the "intermediate" stage, but would have no function in the final product. Another example is that of free-flow agents. Small quantities of these, *e.g.* silicates, may be added to vitamin premixes to ensure that they flow evenly into mixing vessels so that the vitamins are evenly distributed throughout the final product. Release agents are used to prevent food from sticking to a mould or, perhaps, slicing equipment. Again, these uses are to ensure ease and consistency throughout the production process but the additives have no function in the final food. This, then, is the essential technical difference between a processing aid and an additive. Processing aids do not have to be declared on the label.

Regulation 1333/2008 recognises 26 different classes of use for additives in the EU. These are:

Acids, acidity regulators, anticaking agents, antioxidants, antifoaming agents, bulking agents, colours, carriers, emulsifiers, emulsifying salts, firming agents, flavour enhancers, flour treatment agents, foaming agents, gelling agents, glazing agents, humectants, modified starches, packaging gases, preservatives, propellants, raising agents, sequestrants, stabilisers, sweeteners and thickeners.

The Regulation requires that additives may be used only if:

- they do not pose a safety concern to the health of the consumer at the level of use proposed;
- there is a reasonable technological need;
- and their use does not mislead the consumer.

The Regulation clarifies that food additives must serve one of the following purposes:

- preserving nutritional quality of the food;
- providing necessary ingredients or constituents of foods manufactured for groups of consumers with special dietary needs;
- enhancing the keeping quality or stability of a food;
- improving the organoleptic properties, provided that the nature, substance or quality of the food is not changed in such a way as to mislead the consumer;
- aiding in the manufacture, processing, preparation, treatment, packaging, transport or storage of food, including food additives, food enzymes and food flavourings, provided that the food additive is not used to disguise the effects of faulty raw materials or of any undesirable practices or techniques, including unhygienic processes or techniques, during the course of any such activities.

These are similar to the principles enshrined in the Codex Alimentarius, the joint FAO/WHO body responsible for international standards in food.

The harmonisation of food legislation in Europe that began in the 1980s was a prerequisite for trade in the Single Market as differences in national legislation constituted barriers to trade. The legislative position is set out in detail in a later chapter but it is important to appreciate that the development of a new raft of additives legislation was not indicative of the absence of controls before that time, but a recognition that differences in national approaches throughout the Member States were not conducive to the free movement of goods within a single economic entity.

The primary aim of the food industry is to meet consumer needs by providing a wide range of safe, wholesome, nutritious and attractive products at affordable prices all year round, responding to consumer requirements for quality, convenience, price and variety. It would be impossible to do this without the use of food additives. They are essential in the battery of tools used by the food manufacturer to convert agricultural raw materials into products that are safe, stable, of consistent quality and readily prepared and consumed.

Different types of additives are used for different purposes, though many individual additives perform more than one function. For the purposes of both classification and regulation in the EU, they are grouped according to their primary function. The main groupings or classes of additives listed above are explained in the following sections; together with their functions and some examples of their use.

1.2   Preservatives

Preservatives are probably the single most important class of additives, as they play an important role in the safety of the food supply. Despite this fact, any chemical used to counteract the perishability of food raw materials has often been perceived as suspect, and any food containing a preservative has been considered inferior or unsafe. Yet the use of chemical preservatives such as sulfur dioxide, sulfites and sodium nitrate is but a continuation of the age-old practices of using salt, nitrate and spices to preserve perishable foods in the days before refrigeration and modern processing techniques. All food raw materials are subject to biochemical processes and microbiological action that limit their keeping qualities. Preservatives are used to extend the shelflife of certain products and ensure their safety through that extended period. Most importantly, they retard bacterial degradation that can lead to the production of toxins and cause food poisoning. Thus, they offer a clear consumer benefit in keeping food safe over a shelflife that may be extended by the demands of modern lifestyles including infrequent bulk shopping expeditions. The continued perception of preservatives as undesirable, to which the many labels protesting "no artificial preservatives" testify, is therefore an unfortunate consumer misapprehension.

1.3 Antioxidants

Antioxidants reduce the oxidative deterioration that leads to rancidity, loss of flavour, colour and nutritive value of foodstuffs. Many fats, oils, flavouring substances, vitamins and colours react with oxygen when exposed to air. The rate of deterioration can vary considerably and is influenced by the presence of natural antioxidants and other components, availability of oxygen, and sensitivity of the substance to oxidation, temperature and light, for example. Oxidation can be avoided, or retarded, by a number of means, such as replacing air by inert packaging gases, removal of oxygen with enzymes such as glucose oxidase, incorporation of UV-absorbing substances in transparent packaging materials, cooling, and the use of sequestering agents. These may not be possible in all cases, or sufficient for an adequate shelflife for some foods. Some antioxidants remove oxygen by self-oxidation, *e.g.* ascorbic acid, whilst others interfere in the mechanism of oxidation, *e.g.* tocopherols, gallic acid esters, BHA and BHT. All have specific properties, making them more effective in some applications than in others. Often a combination of two or more antioxidants is more effective than any one alone because they are synergistic. The presence of sequestering agents, such as citric acid, may also have a synergistic effect, by reducing the availability of metallic ions that may catalyse oxidation reactions. The use of the powerful synthetic antioxidants BHA, BHT and the gallic acid esters is very restricted (though an EFSA opinion published in March 2012 increased the ADI of BHT from 0–0.05 mg/kg bw/day established by the SCF in 1987 to 0.25 mg/kg bw/day, opening up the possibility of extension to existing authorisations). Tocopherols, which can be either natural or synthetic, are less restricted but are less effective in the protection of processed foods. Antioxidants cannot restore oxidised foods; they can only retard the oxidation process. As oxidation is a chain-reaction process, it needs to be retarded as early as possible. The most effective use of antioxidants is therefore in fats and oils used in the manufacturing process.

1.4   Emulsifiers and Stabilisers

The purpose of emulsifiers and stabilisers is to facilitate the mixing together of ingredients that normally would not mix, namely oils or fats and water. This mixture of oil droplets in water, or water droplets in oils is an emulsion that is created by a vigorous mixing action and sustained by an emulsifier, and often a stabiliser. These additives are essential in the production of mayonnaise, chocolate products and fat spreads. The production of reduced fat spreads, which have made a significant contribution to consumer choice and dietary change, would be impossible without emulsifiers and stabilisers. Anyone who has ever made an emulsified sauce, such as mayonnaise or hollandaise, will appreciate the benefits of this technology – still more so those who have failed in the technique and ended up with an expensive mess of curdled ingredients.

1.5   Colours

Colours are used to enhance the visual properties of foods. Their use is particularly controversial, partly because colour is perceived by some as a means of deceiving the consumer about the nature of the food, but also because some of the most brightly coloured products are those aimed at children. This controversy reached a new height in 2007 with the publication of the so-called "Southampton Study"[3] (see following section on Intolerance). As with all additives, their use is strictly controlled and permitted only where a case of need is proven, *e.g.* to restore colour that is lost in processing, such as in canning or heat treatment; to ensure consistency of colour; and for visual decoration. The use of colour in food has a long and noble tradition in the UK. Medieval cooks were particularly fond of it. The brilliant yellow of saffron (from which Saffron Walden derives its name) and the reddish hue of saunders (powdered sandalwood) were used along with green spinach and parsley juice to colour soups in stripes or to give marbleised effects.[4] So whilst adding colour to food may appear to some as an unnecessary cosmetic that is not in the consumer's interests, there can be no doubt that the judicious use of colour enhances the attractiveness of many foods. Some years ago a number of retailers tried introducing ranges of canned vegetables and fruits such as strawberries and peas without adding back the colour leached out by the heat processing. They were still trying to dispose of the unsold returns several years later! Colour is important in consumer perception of food and often denotes a specific flavour. Thus, strawberry flavour is expected to be red and orange flavour orange-coloured. Consumer expectation is therefore a legitimate reason for adding colour.

1.6   Sweeteners

Sweeteners perform an obvious function. They come in two basic types – "bulk" and "intense", and are permitted in foods that are either energy reduced or have no added sugar. They are also sold direct to consumers as "table-top" sweeteners – well known to dieters and the diabetic. Intense sweeteners, such as aspartame, saccharin, acesulfame K, and more recently sucralose and steviol glycosides have, as their name suggests, a very high sweetening property, variable from type to type but generally several magnitudes greater than that of sucrose, (for example, aspartame is approximately 200 times sweeter than sugar, weight for weight; saccharin 300–500 times; and acesulfame K 130–200 times.) Bulk sweeteners, such as sorbitol, isomalt and maltitol are less sweet, but provide volume and hence mouthfeel. Both types of sweetener are useful in low-calorie products, and are increasingly sought after by many consumers, and for special dietary products such as those for diabetics. The absence of sucrose also lowers the cariogenic properties of the product.

1.7 Flavour Enhancers

Flavour enhancers are substances that have no pronounced flavour or taste of their own but that bring out and improve the flavours in the foods to which they are added. Although salt has a distinctive taste of its own and is not classed as a food additive, it is in fact the most widely used flavour enhancer. The next best known is glutamic acid and its salts, most commonly found in the form of monosodium glutamate, which has been used for several centuries in the Far East as a condiment in savoury products. It is a normal constituent of all proteins, an essential amino acid and present in the body. It is naturally present in a number of foodstuffs including cheese, tomatoes and edible seaweed. Monosodium glutamate (MSG, E 621) has attracted adverse attention in the past for causing an intolerance reaction but this was never confirmed by scientific studies.

Some sweeteners have also been found to have flavour-enhancing properties and have been authorised for use as such. For example, neohesperidine DC (E 959) can enhance the flavour of meat products and margarine, and acesulfame K, aspartame and thaumatin are used to enhance the flavour of chewing gum and desserts.

1.8 Flavourings

Although flavour enhancers are categorised as additives, flavourings are technologically different and regulated separately; even though they are often considered by the general public to be the same thing. Flavourings are defined as imparting odour and/or taste to foods and are generally used in the form of mixtures of a number of flavouring preparations and defined chemical substances. Some 2800 such substances have been identified and, under Regulation 2232/96, were placed on a Register and assessed for safety by EFSA, a process which is still ongoing. The Union List, which converts this Register into the list of approved flavouring substances for use in the EU to the exclusion of all others, was adopted in October 2012.[5]

Flavourings are subject to the same risk assessment and authorisation procedure as additives and enzymes as laid down in Regulation 1331/2008. Regulation 1334/2008 lays down the rules under which flavouring substances may be used. Edible substances and products intended to be consumed as such, or substances that have exclusively a sweet, sour or salty taste, i.e ordinary food ingredients such as sugar, lemon juice, vinegar or salt are excluded from this Regulation. Flavourings are divided into the following categories; flavouring substances, flavouring preparations, thermal process preparations, smoke flavourings, flavour precursors or other flavourings or mixtures. A flavouring substance is a single chemical with flavouring properties, for example ethyl vanillin, whereas a flavouring preparation is a product that may contain several individual flavouring substances, and is obtained from food, for example vanilla extract. Because of the complexity of the flavouring used in a food, labels generally indicate simply "flavourings". This is all that is legally required, as to list every individual substance would often be extremely lengthy and virtually incomprehensible to the consumer, although the manufacturer may be more specific if he wishes. Any flavourings labelled as "natural" must meet the definition in article 3 of Regulation 1134/2008. In the UK, the Food Standards Agency has issued guidance on the use of the word "natural" in product labelling.

As with additives, some flavourings are sold direct to the consumer for domestic, culinary use. Vanilla and peppermint are amongst the best known, as well as the popular brandy and rum essences. Anyone who has ever overdone the amount of flavouring in a home-made cake or a batch of peppermint creams will appreciate the minute quantities in which they are used. Similarly, in commercial manufacture, the quantity of flavouring used is extremely small in relation to that of other ingredients. Most flavourings are developed from substances naturally present in foods. Citrus and orange oils, for example, are amongst the most common natural source materials used in flavouring preparations and substances.

1.9   Other Additives

For regulatory purposes, colours and sweeteners are very specific, well-defined classes of additives and, because of the nature of their function, have been subject to specific legislation. All other classes of additive fall under the general heading of "miscellaneous", though the regulatory distinction is more blurred now that the specific additive Directives of the 1990s have been amalgamated into the additives Regulation, 1333/2008, and all authorisations are listed under the new Food Categorisation System (FCS) in Annex II of the Regulation. In addition to the larger groups of additives such as preservatives, emulsifiers and stabilisers mentioned above, there are other categories within this more general grouping – namely acids, acidity regulators, anticaking agents, antifoaming agents, bulking agents, carriers, glazing agents, humectants, raising agents, sequestrants and thickeners.

The function of most of these is obvious from the name, with the possible exception of sequestrants. These are substances that form chemical complexes with metallic ions. They are not widely used and it is a class of additives rarely seen on a food label, but they perform an important role as they prevent the metallic ions, which may arise from other ingredients or from water, from catalysing oxidation reactions or causing curdling in dairy products. Thickeners, on the other hand, are amongst the most commonly used additives, as they exert an effect on the texture and viscosity of food and drinks products. Much as various types of flour are used extensively in the kitchen to thicken sauces, soups, stews and various dishes with a high liquid content, most commercial thickeners are starch or gum based and serve much the same purpose.

One class of additive that has no domestic equivalent is that of packaging gases. These are the natural atmospheric gases now widely used in certain types of prepacked products, such as meat, fish and seafood, fresh pastas and ready-prepared vegetables found on the chilled food counters in sealed containers. The "headspace" of the container is filled with one, or a combination of the gases, depending on the product, to replace the air and modify the atmosphere within the pack to help retard bacteriological deterioration, which would occur under normal atmospheric conditions – hence the term "packaged in a modified atmosphere". Arguably, the gases do not have an additive function as they are not detectable in the food itself and function only to preserve the food for longer in its packaged state, but for regulatory purposes they were deemed to be additives and must therefore be labelled. Carbon dioxide will, of course, also be familiar as an ingredient in many fizzy drinks – an illustration of the many different functions and uses of additives.

In the EU, enzymes, with the exception of lysozyme and invertase, are not included in Regulation 1333/2008 but come under a separate Regulation, 1332/2008. However, the general principles relating to their use are those of 1331/2008. They will eventually be included in a Community list of enzymes permitted for use in foods and food processing, which remains under development, as all food enzymes are being evaluated by EFSA in the first full harmonisation measure for enzymes in the EU.

1.10 Safety of Additives

The safety of all food additives, whether of natural origin or synthetically produced, is rigorously tested and periodically reassessed. In the EU this process is carried out by expert committees of the European Food Safety Authority (EFSA) Panel on Food Additives and Nutrient Sources Added to Food, (the ANS Panel). Only additives that have been evaluated and approved are given an "E" number; thus the number is an indication of European safety approval, as well as a short code for the name of the additive. As part of the regular re-evaluation process EFSA issued an opinion in 2004 on the safety of the parabens (E 214–219). As a result, Directive 2006/52 deleted the propyl p-hydroxybenzoate and its sodium salt. Similarly, in a re-evaluation of Red 2G,

E 128, in 2007 EFSA decided that there was a safety concern and this was subsequently deleted from the list of permitted additives.

Individual countries also have similar procedures and at international level, the Joint Expert Committee on Food Additives (JECFA), which advises the UN's Food and Agriculture Organisation (FAO) and World Health Organisation (WHO) Codex Alimentarius Commission, evaluates additives. This latter has become increasingly important in recent years as the World Trade Organisation arrangements specify that Codex standards will apply in any dispute over sanitary and phytosanitary standards, *i.e.* the safety and composition of foods. For this reason the Codex General Standard for Food Additives (GSFA) was adopted to recommend usage levels of food additives in all products traded internationally.

In evaluating an additive, EFSA, JECFA and other national agencies allocate an "Acceptable Daily Intake" (ADI), the amount of the substance that the panel considers may be safely consumed, daily, throughout a lifetime. This assessment is used to set the maximum amount of a particular additive (or chemically related group of additives) permitted in a specific food, either as a specified number of grams or milligrams per kilogram or litre of the food or, if the ADI is very high or "nonspecified", at *quantum satis*, *i.e.* as much as is needed to achieve the required technological effect, according to Good Manufacturing Practice.

In establishing the ADI, a safety factor is always built in, usually 100-fold, to ensure that intake of any additive is unlikely to exceed an amount that is anywhere near toxicologically harmful. To ensure that consumers are not exceeding the ADI by consuming too much of or too many products containing a particular additive, EU legislation requires that intake studies be carried out to assess any changes in consumption patterns. This process is elaborated in Chapter 2 but it is worth noting that most additives are self-limiting and there is rarely any benefit to the product in using more than is necessary. It is good manufacturing practice to limit the amount of additive used.

1.11 Intolerance

"Additives" have often been blamed for causing intolerance or allergic reactions, especially hyperactivity in children. Whilst there is no doubt that certain foods and food ingredients, including additives, are responsible for intolerance reactions in some people, this generic accusation is a wild exaggeration. Though many people believe themselves to be sensitive to certain ingredients, the true prevalence of intolerance to foods has been shown to be about 2% in adults and up to 20% in children, and for food additives from 0.01 to 0.23%. The substantial overestimation of such reactions by the general public probably has its origins in the adverse media coverage and antiadditives campaigning of the 1980s, when popular belief was that additives were responsible for harmful behavioural effects and hyperactivity was attributed solely to the consumption of artificial colours. The Food Standards Agency

(FSA) in the UK commissioned a study by Southampton University to investigate the impact of consuming drinks containing a mix of artificial colours (tartrazine (E 102), quinoline yellow (E 104), sunset yellow (E 110), ponceau 4R(E 124), carmoisine (E 122) and allura red (E 128)) and sodium benzoate on young children. Whilst the results were not clear cut, the FSA decided that it should advise parents of young children who showed signs of hyperactivity or Attention Deficit Hyperactivity Disorder (ADHD) that avoiding the colours used in the study might be beneficial and urged the European Commission to ban them. Acting on a request from the European Commission, EFSA reviewed the study and concluded that "the findings of the McCann *et al.* study could not be used as a basis for altering the ADIs of the respective food colours or sodium benzoate".[6] Despite this conclusion, the publicity accorded the study led the European institutions (which include the European Parliament as well as the European Commission and Member States) to include a requirement in Regulation 1333/2008 that the following warning statement should appear on the label of products containing the colours that were the subject of the study: "[name or E number of the colour] may have an adverse effect on activity or attention in children". In subsequent reviews of the safety assessments of the colours, carried out as part of the rolling programme of safety reviews of all food additives under Regulation 1333/2008, EFSA concluded that the ADIs of three colours, ponceau 4R, quinoline yellow and sunset yellow should be reduced. This was based on exposure assessments that suggested the ADIs of these colours could be exceeded in certain population groups and had nothing to do with the McCann *et al.* study or ADHD. Action has now been taken to consider the uses of the colours concerned and authorisations amended as necessary to ensure the ADIs are not exceeded.

Food intolerance, and especially allergy, is frequently under the spotlight because of the seemingly growing occurrence of severe allergic reactions, particularly to peanuts. Since the mid-1990s, there have been a number of widely reported incidents, including several tragic deaths as a result of anaphylactic shock, a severe allergic reaction to specific proteins, most commonly those found in tree nuts and peanuts and a small number of other foods, including milk, wheat, eggs, soya, fish and shellfish. The reasons for such reactions are not yet fully understood and are the subject of a great deal of research, as are the causes of this apparently growing problem. The need to address the issue and do everything possible to assist the small but significant number of people affected by this most severe form of allergy caused the European Commission to task its former Scientific Committee for Food (SCF) with identifying the scope of the problem and the foods and ingredients associated with it. The 1997 SCF report estimated that the prevalence of intolerance to food additives was 0.026% of the European population,[7] which compares to the prevalence of adverse reactions to cows' milk of 1 to 3%. EFSA has accepted a mandate to review the state of play of the current knowledge in the field of allergens, including prevalence figures. This review is expected to be finalised by the end of 2012.

The most commonly observed reaction from food additives is to sulfur dioxide (E220) and sulfites (E 221–E228), especially in asthma sufferers. Within the EU, allergens have had to be declared on the ingredient label since 2006 and this requirement has been extended to include 14 specific allergens in the Food Information Regulation, 1169/2011.

1.12 "Clean Labels"

The growing demand from consumers for products with ingredients lists that contain no additives is driving much product development in Europe. The term "clean label" is used to refer to this trend. A number of ingredients are now being manufactured that claim to fulfil this aim, particularly to replace colour additives. Among antioxidants, rosemary extract has been safety assessed, approved and permitted as an additive (E 392), but there are also grape seed, chestnut and olive leaf extracts. Beetroot extract and grape skin extracts are approved and permitted as additives (E 162 and E 163) and there is an increasing trend towards the use of spinach, pumpkin, nettle and spirulina extracts, establishing a category of colouring foodstuffs. The European Commission and Member States are looking into these uses and are expected to finalise guidance towards the end of 2012. This will clarify the borderline on the "selective extraction" test set out in Regulation 1333/2008, and will distinguish between an ordinary food ingredient and an unauthorised (illegal) food additive. Such ingredients are not regulated; however, the increasing use of novel extracts may bring materials under the ambit of the Novel Foods and Novel Food Ingredients Regulations 258/97. This is a fast growing area of ingredient development and its future in terms of both ingredients and legislation promises much.

1.13 Conclusion

Never has the range and choice of foods available in shops been so great, but then neither have the demands made upon the food industry. Legislation ensures that our food is safe and labelled to provide all the information the consumer needs to choose their selection from the wide range of options available. Food additives have enabled the industry to provide foods that meet consumer demands for taste, variety, convenience and safety.

References

1. M. Hanssen, *E for Additives – The Complete E Number Guide*, Thorsons, Wellingborough, 1984.
2. Regulation 1333/2008 of the European Parliament and of the Council of 16th December 2008 on food additives L354, 31.12.2008, 16–33.
3. D. McCann *et al.*, *Lancet*, 2007, **370**, 1560–1567.

4. M. McKendry, *Seven Hundred Years of English Cooking*, Treasure Press, London, 1973.
5. Commission Implementing Regulation (EU) No 872/2012 of 1 October 2012 adopting the list of flavouring substances provided for by Regulation (EC) No 2232/96 of the European Parliament and of the Council, introducing it in Annex I to Regulation (EC) No 1334/2008 of the European Parliament and of the Council and repealing Commission Regulation (EC) No 1565/2000 and Commission Decision 1999/217/EC, *Official Journal* L 267, 2 October 2012, **55**, 1–161.
6. Assessment of the results of the study by McCann *et al.*, (2007) on the effect of some colours and sodium benzoate on children's behaviour, Scientific Opinion of the Panel on Food Additives, Flavourings, Processing Aids and Food Contact Materials (AFC), (Question No EFSA-Q-2007-171), Adopted on 7 March 2008, *The EFSA Journal*, 2008, **660**, 1–54.
7. Scientific Committee for Food, *Report on Adverse Reaction to Food and Food Ingredients*, Luxembourg, 1997.

CHAPTER 2

Safety of Food Additives in Europe

SUSAN M BARLOW

Harrington House, 8 Harrington Road, Brighton, East Sussex BN1 6RE, UK
E-mail: suebarlow@mistral.co.uk

2.1 Introduction

The objective of European Union (EU) legislation on food additives is to ensure protection of public health within a harmonised EU internal food market. The legislation on food additives has been developed following the approach laid down by the European Commission in 1985.[1] This approach limited the requirement for legislation to those areas that were justified by the need to protect public health, to provide consumers with information and protection in matters other than health, to ensure fair trading and to provide for the necessary public controls. This chapter focuses on the processes that are in place to ensure the safety of food additives covered by EU legislation.

2.2 Definition of a Food Additive

In the EU, a food additive is defined in law as:

"Any substance not normally consumed as a food in itself and not normally used as a characteristic ingredient of food, whether or not it has nutritive value, the intentional addition of which to food for a technological purpose in the manufacture, processing, preparation, treatment, packaging, transport

Essential Guide to Food Additives, 4th Edition
Edited by Mike Saltmarsh
© The Royal Society of Chemistry 2013
Published by the Royal Society of Chemistry, www.rsc.org

or storage of such food results, or may be reasonably expected to result, in it or its by-products becoming directly or indirectly a component of such foods."[2]

The definition excludes processing aids, including enzymes and extraction solvents, flavourings, substances added as nutrients, such as vitamins and minerals, and substances migrating from food contact materials (food packaging, utensils, *etc.*) which do not exert a technological function in the food. All substances falling under this definition are simply called food additives in the EU, whereas in the USA, the terms "direct food additive" and "color additive" are used to describe them. The US Food and Drugs Administration (FDA) also uses the term "indirect food additives" to describe pesticide residues in food and substances derived from food contact materials that are present in food. In the EU, pesticides and substances used to make food contact materials are covered by separate legislation. Similarly, extraction solvents and flavourings are also covered by separate EU legislation.

2.3 European Legislation and the Safety Assessment Process

2.3.1 The Framework Legislation

A general framework Regulation (EC) No 1331/2008 establishing a common authorisation procedure for food additives, food enzymes and food flavourings was adopted in December 2008.[3] This replaced an earlier 1988 framework Directive on food additives.[4] Three further Regulations were also adopted in 2008, relating to additives,[5] enzymes,[6] and flavourings and certain food ingredients with flavouring properties.[7] These Regulations lay down the harmonised criteria and requirements for the assessment and authorisation of these substances for food use. In the EU, food additives, food enzymes and food flavourings are collectively termed "food improvement agents".

A later Commission Regulation (EU) No 234/2011[8] elaborates on procedural arrangements for updating the existing lists of substances authorised for use in food in the EU. It sets out the formal procedure for submitting an application to market a new food additive, enzyme, or flavouring and the required content of the dossiers to be submitted in support of an application. Dossiers should contain (*inter alia*) the essential information for risk assessment, including technical information, biological and toxicological data relating to safety, information on proposed uses, normal and maximum use levels, and estimates of dietary exposure.

2.3.2 Legislation on Specific Classes of Additives

In 1994–95, three specific Directives stemming from the 1988 framework Directive[4] were adopted and these currently remain in force. They cover sweeteners,[9] colours,[10] and additives other than sweeteners and colours,[11] the latter more usually being referred to as "miscellaneous additives". These Directives list the individual permitted additives, the general and specific food categories in which each additive is permitted, and where necessary, maximum permitted levels of use. All three Directives also require Member States to set up systems to monitor consumer consumption of additives and, in the case of sweeteners, to establish consumer surveys that will include monitoring of "table-top" sweetener usage.[12] Results of such monitoring must be reported to the Commission and ultimately to the European Parliament.

From mid-2013, the new legislation on food additives will be fully in force. The additives approved in the EU for use in food and their conditions of use have been harmonised into a single list that can be found in Annex II of the general framework Regulation (EC) No 1331/2008,[3] which applies from 1 June 2013. Prior to that date the provisions of the former Directives on specific classes of additives continue to apply. A further list of approved food additives for use in food additives, food enzymes and food flavourings and nutrients (*e.g.* carriers) can be found in Annex III of the general framework Regulation.

2.3.3 The Role of the European Food Safety Authority

Since 2003, the risk assessment of food additives has been undertaken by an agency of the Commission, the European Food Safety Authority (EFSA). Unlike the regulatory systems governing other chemical sectors, such as human and veterinary medicines, and pesticides, which have advisory committees on which representatives of all the EU Member States sit, EFSA is required by its founding Regulation[13] to use independent scientific experts on its Scientific Panels and Committee. These experts do not represent Member State governments and are required to give their own independent views. The creation of this agency ensured that the risk managers (the European Commission and the EU Member States) could receive independent scientific advice on proposed food additives prior to them being considered for authorisation. It also ensures that there is an appropriate separation between risk assessors and risk managers.

EFSA has published opinions on the data requirements for food additives,[14] enzymes[15] and flavourings,[16] which discuss the range of information and toxicity studies that must be submitted in support of an application for these types of substances. Upon receipt of a dossier, EFSA verifies whether the information submitted is suitable for undertaking a risk assessment. If the dossier is considered valid, EFSA examines the data in depth and provides the risk managers with an opinion on whether the substance is safe for consumers, under the proposed uses and conditions of use. EFSA opinions on food additives summarise the available data and provide conclusions on the safety/toxicity of the additive, dietary exposure assessments, including exposure of vulnerable consumer groups, and, where appropriate, establish a health-based guidance value, such as an acceptable daily intake (ADI) for the additive (see Section 2.6.1). All the published opinions on proposed and existing individual food additives can be found on the EFSA website (http://www.efsa.europa.eu/).

Many food additives currently permitted for use in the EU were first evaluated some time ago, some as long ago as the 1970s, by the former EC Scientific Committee on Food (SCF), which provided advice to the Commission in the 25 years prior to the launch of EFSA. In addition to evaluating applications for new food additives, EFSA has also been charged, under Regulation EU 257/2010,[17] with the task of reviewing all permitted food additives that were authorised before 2009 in a re-evaluation programme that will run until 2020.

2.3.4   General Criteria for the Use of Food Additives

The general criteria for use of food additives are set out in Regulation (EC) No 1333/2008.[5] The Regulation stipulates that an additive can be approved only provided it presents no hazard to the health of the consumer at the level of use proposed, that there is a reasonable technological need that cannot be achieved by other economically and technologically practicable means, and that its use does not mislead the consumer. All food additives must also be kept under observation after approval so that they can be re-evaluated if there are changing conditions of use or new scientific information emerges on safety aspects.

2.4   Origin of "E" Numbers

Each permitted additive is assigned an "E" number, signifying that it has been approved as safe for food use and its inclusion in the relevant Directive or Regulation has been agreed by the Member States. Each E number has a separate specification that lays down purity criteria for the additive.[18–20] Labels on processed foods must list additives by their E numbers and/or by their common name.[2] An up-to-date list of authorised substances can be found on the EU database on food additives (available at: https://webgate. ec.europa.eu/sanco_foods/main/?event=display).

2.5   Safety Testing of Food Additives

2.5.1   Toxicological Tests Required

To assess the possible adverse effects of a proposed food additive, it must be subject to toxicological testing. Most commonly, such tests will be laboratory studies, including a range of tests conducted in experimental animals. Occasionally, there may be information from human studies, but it is more likely that such studies only become available after a food additive has been marketed.

There is a scientific consensus on the range of toxicity tests that should be considered as appropriate for the testing of food additives.[21] Not all studies are necessarily required for every additive. If a proposed additive is likely to have a large number of uses and potentially widespread daily exposure, then a full range of toxicity studies will normally be required. However, if a proposed additive will have limited uses, or human exposure can be estimated to be very low (*e.g.* because the substance concerned is very poorly absorbed from the gut), then a smaller range of studies may be sufficient to reach a conclusion on its safety in use.

The range of toxicological tests generally required for a proposed new food additive have been set out by various bodies including the Joint FAO/WHO Expert Committee on Food Additives (JECFA),[21] the EFSA,[22] and the US FDA.[23] While these various guidelines differ in some aspects of detail, the core requirements are very similar.

Toxicological tests are usually conducted to standard protocols, the guidelines for which have been developed and agreed internationally. Studies conducted to the Organisation for Economic Co-operation and Development (OECD) Guidelines[24] or to EU Guidelines,[25–27] the latter being essentially the same as the OECD's, are acceptable for the testing of food additives for applications made to the EU.

2.5.1.1   Absorption, Distribution, Metabolism and Excretion

Studies on absorption, distribution, metabolism and excretion (ADME), also termed toxicokinetics, are usually conducted following single and short-term, repeated dosing in experimental animals. Such studies can greatly aid in the design of subsequent toxicity tests, indicate whether potentially harmful metabolites may be produced, or whether the parent compound or its metabolites may accumulate in the body. They are also helpful for the interpretation of adverse findings in toxicity tests. In the EFSA guidance, toxicokinetic studies are regarded as a core part of the tiered testing strategy, to be conducted at an early stage.[22] Ideally, toxicokinetic information in humans is also desirable, for example, so that toxicokinetic information from the experimental species used in the toxicity tests can be compared with humans, but ethically, such information from humans can only be obtained after an additive has been extensively tested in animals to demonstrate the likely safety of the substance.

2.5.1.2   Acute Toxicity

Few bodies now require acute toxicity testing of food additives, though such data will often exist because they must be provided for occupational safety purposes in manufacturing. Similarly, studies on eye and skin irritation and skin sensitisation may also be conducted for occupational safety assessments, but they are of little use in the safety evaluation of food additives for human consumption. Skin sensitisation, for example, is not predictive of oral sensitisation, for which no validated animal models yet exist.

2.5.1.3   Subchronic Toxicity

A subchronic study in a rodent species (usually rat), and sometimes also in a nonrodent species (dog), is generally required. The duration of a subchronic study is usually 90 days and exposure to the test substance is continuous. The test substance is normally given by the oral route, administered at fixed concentrations in the diet. Sometimes it is given by oral gavage, such that the test substance is delivered as a bolus dose directly into the stomach. For nontoxic substances, inclusion of upper concentrations in the diet that are greater than 5% are not advisable, since such concentrations can cause nutritional disturbances that may then give rise to secondary toxicity. Subchronic studies yield important information on food consumption and body weight, haematology, blood and urine biochemistry (that can provide indications of damage to organs such as the liver and kidney), organ weights, and pathological effects in organs and tissues at the gross, macroscopic and the microscopic levels.

2.5.1.4   Reproductive and Developmental Toxicity

Reproductive and developmental toxicity studies are also usually required. These generally comprise a reproductive study in the rat, and developmental toxicity studies in one or two species. In a reproductive study, the test substance is usually administered continuously in the diet over two generations, so that the reproductive function of the parents and of the offspring exposed to the test compound *in utero* can be assessed. Reproductive studies provide information on male and female fertility, maintenance of pregnancy, birth and lactation, and indicate any adverse effects on survival, growth and development of the offspring. Nowadays such studies are likely to include not only assessment of postnatal physical development of the offspring but also measures of motor and behavioural development. This is regarded as very important, since critical aspects of the development of the reproductive system in rats occur in the late prenatal and early postnatal period, a developmental window in which there may be particular vulnerability to endocrine-mediated adverse effects.

In developmental toxicity studies (formerly known as teratology studies) the growth and development of the embryo and fetus is assessed, with emphasis on embryonic and fetal survival, fetal weight and the occurrence of any malformations. The dosing period for such studies is throughout embryogenesis and until just prior to birth. This is in order to cover important periods of brain and reproductive system development, which continue beyond the shorter period of organogenesis for other systems. Dosing is normally via the diet or by gavage.

2.5.1.5   Chronic Toxicity/Carcinogenicity

Chronic toxicity/carcinogenicity studies in two species, usually rat and mouse, are normally required for many food additives. Dosing commences when the animals are in the juvenile period of rapid growth, at about 6 weeks of age, and continues for most of the animal's lifetime (*e.g.* 18 months for mice and 24 months for rats). The test substance is usually administered in the diet. The emphasis in these studies is on discovery of any effects on body weight, organ weight or pathological changes in tissues and organs. Examination of haematological and clinical chemistry parameters may also be included in satellite groups killed at intervals before the termination of the study.

2.5.1.6   Genotoxicity

Genotoxicity studies assess the ability of a substance to interfere with DNA by induction of gene mutations, chromosome aberrations or other forms of DNA damage. Such effects are of significance because they indicate the potential for carcinogenic effects, induction of heritable mutations in germ cells, or other adverse health consequences of genotoxicity. Usually, two or three *in vitro* tests covering the endpoints of gene mutation, chromosome aberration and aneu-ploidy are required, using bacterial systems such as *Salmonella typhimurium* and mammalian cells in culture from rodents or human lymphocytes. *In vivo* follow-up tests may also be required, especially if genotoxic effects are obtained in any of the *in vitro* studies. This is in order that the genotoxic potential of the substance in the context of its *in vivo* metabolism and kinetics can be assessed. Substances that are genotoxic *in vitro* but not *in vivo* (*e.g.* because they are readily broken down into nongenotoxic compounds) are not generally regarded as hazardous to humans. For food additives, carcinogenicity studies (see above) will also often be available to provide data to be considered alongside the information on genotoxicity.

2.5.1.7   Other Studies

The core studies are designed to give clear information on the nature of any toxicity and to identify no-observed-adverse-effect levels (NOAELs) for any toxicological effects observed. However, for some aspects of toxicity, the core studies are designed only to indicate a potential problem and further studies may be required to properly elucidate these. Depending on the findings in the core tests, further special studies may sometimes be needed, for example, to clarify the mode or mechanism of action for any toxicity in order to determine its relevance to humans, or to better define the (cellular, subcellular, biochemical) doses that are without any adverse effect. Similarly, if the core studies indicate, for example, that there may be effects on the immune, nervous or endocrine systems, then further special studies designed to answer specific questions on these aspects may be required.

2.5.2   Outcomes of Toxicity Tests

The toxicity of most food additives is generally low in comparison with pesticides, drugs and some industrial chemicals. Some food additives show no toxicity (*e.g.* modified starches and celluloses), even at the highest doses administered that are compatible with not disturbing nutritional balance. More common is to find that there is some manifestation of toxicity, particularly at the highest dose administered. Indeed, standard protocols for toxicity testing require that the highest dose be chosen with the aim to induce toxicity but not death or severe suffering.

Typical of the types of effects observed in animals in subchronic and chronic toxicity studies on food additives are reductions in body weight gain, with or without accompanying reductions in food consumption. Effects on the liver and kidney are also not uncommon because these are the major organs of metabolism and elimination, so are often exposed to the highest concentrations of the additive and its metabolites.

In reproductive and developmental toxicity studies, effects on the offspring, such as reduced survival, reductions in body weight and postnatal growth, physical and behavioural abnormalities, or other changes are assessed. If adverse effects are observed, they need to be considered in the light of whether they are a primary effect on the offspring or secondary effects due to maternal toxicity. However, concluding on whether an effect on the offspring is secondary to maternal toxicity can often be difficult, particularly as the top dose used in a developmental study should cause some maternal toxicity.

In chronic toxicity/carcinogenicity studies, provided the highest dose does not exceed the maximum tolerated dose of a substance (*e.g.* a reduction in body weight not exceeding 10% in comparison with controls), the general toxicity does not usually interfere unduly with interpretation of the results with respect to carcinogenicity.

The interpretation of genotoxicity and carcinogenicity studies is of special significance for food additives. Any substance that is genotoxic *in vivo* would not be regarded as acceptable for use as a food additive, since such effects may be without a threshold and thus could occur at very low daily exposures over a lifetime. If a substance is genotoxic *in vivo* it often produces tumours in carcinogenicity bioassays.

Substances that are nongenotoxic *in vivo*, may also show evidence of carcinogenicity in lifetime rodent studies (*e.g.* the sweetener, sodium saccharin).[28] Such substances may act by inducing tissue damage and necrosis, resulting in enhanced cell division during repair, and triggering tumour development, but the effects have a threshold. Provided a dose causing no tissue damage can be identified, then such substances may be acceptable as food additives.

2.5.3 Relevance of Effects Observed in Animals for Human Risk Assessment

In reviewing the entire database, expert judgment has to be made about which effects are adverse and which are not. For example, the antioxidant, butylated hydroxyanisole (BHA) causes forestomach tumours in the rat when fed at 1 and 2% in the diet. This is due to prolonged stimulation of the forestomach epithelium, resulting in hyperplasia, which is also seen at a lower dose of 0.5% in the diet but not at 0.125% (equivalent to 62.5 mg/kg body weight per day) in the diet.[28] As a currently permitted additive, BHA has been reviewed by EFSA who commented that forestomach hyperplasia may no longer be considered relevant for human risk assessment.[29]

Similarly, the feeding of large amounts of poorly absorbed materials, such as polyol sweeteners, to rats is known to cause caecal enlargement, disturb calcium homeostasis, cause pelvic nephrocalcinosis and perhaps result in the development of adrenal phaeochromocytomas.[30] However, this would not be taken as indicative of the same adverse effects occurring in humans if the additive is used in small amounts. On the other hand, the feeding of large amounts of poorly absorbed bulk sweeteners, such as the polyols, also can cause an osmotic diarrhoea, an effect that is transient but that also occurs in humans. This effect is taken into account in deciding in what foods such additives can be used and in setting maximum levels of use.[30] There are also effects that occur in rodents that are not of significance for man, such as kidney damage and tumours via a mechanism involving α-2μ-globulin, a protein formed only in male rat liver, which binds with certain hydrocarbons and accumulates in the kidney.[31]

Effects may also be observed that are not regarded as being of toxicological significance, such as staining of tissues when high amounts of colours are fed to animals, or increases in liver weight and liver enzyme induction in response to metabolic overload when high amounts of some substances are fed. Similarly, sporadic statistically significant changes in biochemical or haematological parameters, which are inevitable in any series of repeat-dosing studies in which a large number of parameters are investigated, may not be considered relevant for human risk assessment if they are not dose related and/or are not accompanied by corroborating pathological changes.

2.6   Risk Assessment of Food Additives

Risk assessment of food additives (sometimes termed "safety assessment") has developed along similar lines in the EU and in the wider international scientific community. The main international body which has addressed the issue of food additive safety is the JECFA. This Committee advises the Codex Alimentarius Commission, which develops harmonised international food standards, guidelines and codes of practice to protect consumer health and ensure fair trade practices in food trade worldwide. The JECFA was set up in 1956 and over the years has drawn on expertise from around the world for its changing membership. During the first 5 years of its existence the JECFA set out principles for the assessment of food additives, which have been collated and updated in subsequent years.[21] The general principles of the JECFA approach have been widely adopted by other national and international bodies, including EFSA. Both JECFA and EFSA evaluate food additives, often reaching similar conclusions, but occasionally their views may differ, for example, in the precise numerical value of an ADI, or because evaluations were done at a different time and the available toxicological database was not identical. In the EU, the opinions of EFSA take precedence over those of JECFA.

2.6.1 Derivation of an Acceptable Daily Intake

Evaluations of a food additive are usually based on an extensive series of toxicological tests, as described earlier. From these, any toxic effects are identified, together with an examination of their dose–response relationships, including the dose(s) at which it does not cause any adverse effects (NOAEL). All this information is then used to set a health-based guidance value, known as the Acceptable Daily Intake (ADI).

The ADI has been defined by JECFA as an estimate of:

"the amount of a food additive, expressed on a body weight basis, that can be ingested daily over a lifetime without appreciable health risk".[21]

A numerical ADI is derived by selecting the NOAEL from the critical study (or studies), amongst the range of toxicity studies available. The critical study is often a chronic toxicity study or a reproductive study because of the long periods of administration and the fact that such studies cover a number of critical periods in the lifetime. The critical study is usually (though not always) that giving the lowest NOAEL among all the studies for an effect that is considered potentially relevant to humans (see Section 2.5.3). If the potential relevance to humans is unclear, then the default is to consider that it is relevant. The NOAEL from the critical study is then divided by an uncertainty factor, sometimes termed a safety factor, to give the ADI. The uncertainty factor most commonly applied is a default factor of 100, comprising a factor of 10 to take account of possible interspecies differences when extrapolating from animals to humans and a further factor of 10 to take account of possible interindividual variability between humans. The ADI is expressed as a range from 0 to an upper limit, in mg/kg body weight.

Occasionally, the ADI may be based on human data, in which case there is no need to include an uncertainty factor for interspecies differences, and so a default factor of 10 rather than 100 is applied. Such is the case, for example, with the colour Erythrosine. This affects the human and rat thyroid, ultimately causing tumours in the rat due to excessive production of thyroid-stimulating hormone (TSH). Because of well-known differences between rat and human thyroid physiology, tumour production is not considered likely in humans, but increases in circulating thyroid hormone levels can be induced in humans by Erythrosine. Its mechanism of action is well understood and a no-effect level for increases in thyroid hormone levels in humans has been established and used, in conjunction with a 10-fold rather than a 100-fold uncertainty factor, to set the ADI.[32] Human data have also been central to the re-evaluation of other colours, *i.e.* Tartrazine (E102), Quinoline Yellow (E104), Sunset Yellow FCF (E110), Ponceau 4R (E124), Allura Red AC (E129), and Carmoisine (E122). EFSA concluded that a study on the effect of ingestion of mixtures of these colours on children's behaviour provided limited evidence of a small effect on activity and attention.[33] As a result, the European Parliament required foods containing these colours to carry a specific labelling to that effect.[2]

For some food additives evaluated in the past, an "ADI not specified" has been allocated. This is because a number of additives are of very low toxicity and no toxic effects were seen during animal testing when large amounts were given in the diet. Examples of such additives are ones that are the same as normal food ingredients (*e.g.* citric acid) or human metabolites (*e.g.* lactic acid).

For other additives, neither a numerical ADI nor an "ADI not specified" is allocated. Instead, the additive may be judged to have no safety concerns given the proposed uses and proposed levels of use. In these cases, there may be insufficient data to set a numerical ADI (*e.g.* one particular type of study may not be available), or they may be very poorly absorbed, or their proposed use levels may be very low. On the basis of the information available, they are considered not to pose a hazard to health. In such cases, EFSA may reach a conclusion on safety by comparing the highest potential exposures with the NOAELs from the toxicity studies in order to estimate whether there is an adequate margin of safety. Examples of additives considered to have no safety concerns, but without a numerical ADI, are natamycin, a fungicide used only for the surface treatment of some cheeses and dried, cured sausage,[34] and gum acacia modified with octenyl succinic anhydride, used as an emulsifier for flavor-oil preparations that are then added to foods in small amounts.[35]

Structurally related additives, with a common mechanism of action or effect, may be assigned a Group ADI. The summed exposures to all the additives in a related group should not exceed the Group ADI.

2.6.2   Comparison of Exposure to Food Additives in the Diet with ADIs

2.6.2.1   Methods for Estimating Exposure to Food Additives

To assess the health significance, if any, of exposure to food additives present in the diet, the ADI is compared to average and high consumption estimates in the population. Often, exposure also needs to be estimated for particular subgroups of the population, *e.g.* sweetener exposure in consumers with diabetes, exposure to colours in children, or exposure of infants to additives used in infant formula. There are a number of important practical issues in estimating dietary exposure to food additives in order to obtain reliable estimates of average and high consumption, especially for various subgroups of the population with differing dietary habits.[36] It has therefore been usual for dietary exposures to food additives to be estimated initially using relatively crude approaches and for these to be further refined if necessary.

A very rough estimate of dietary exposure to a food additive on a population-wide basis can been made by dividing the total weight of a food additive made annually, or the disappearance annually into the food chain of a food additive, by the number of individuals in the population as a whole.

However, annual per capita consumption figures so generated may be misleading in that they usually do not take account of imports and exports of food and assume that all food sold is consumed. More critically, it is likely to considerably underestimate the actual exposure of individuals because it assumes consumption is even across the entire population, which is rarely the case. Even if the estimate is made assuming that only 10% of the population consumes a particular additive, this may still underestimate exposure in cases where consumers are loyal to a particular brand of food in which the additive is used. Such per capita estimates are rarely useful for providing reassurance that a health-based guidance value is not being exceeded.[37]

An initial screening method that was previously used in the EU is known as the Budget Method.[38] It relies on assumptions regarding physiological requirements for energy and liquid and on energy density of foods, instead of detailed food consumption surveys. It assumes that all foods contributing to energy intake and all beverages contributing to liquid intake will contain the additive at the maximum permitted use levels. The resulting exposure estimation is clearly an overestimate, but if such an estimate is below the ADI for the food additive concerned, then no further refinement of the exposure estimate may be necessary.

A more refined method is to use surveys of food consumption that are representative of the population as a whole, or of subpopulations, and which provide information on average and high consumption of a wide range of foods. It is then assumed that all foods that may contain any particular additive do, and that they do so at the maximum level permitted or maximum level needed to achieve the desired technological effect. Combining these two types of information results in estimated exposures for average and high consumers that are thought to be conservative in that the assumptions made about food additive content are likely to overestimate actual exposure.[39]

EFSA now requires that estimates of dietary exposure to food additives are obtained by combining proposed uses and use levels with extensive information on food intake contained in the EU Comprehensive Food Consumption Database, which has been compiled using information supplied by EU Member States.[22,40]

Since most food additives are regularly ingested and not of any direct health benefit to the consumer, it is important to make comparisons of exposures with ADIs, not only for average consumers but also for high consumers, to ensure they are protected. Different regulatory authorities may use differing cut-off points to define "high" consumers. None use information from the most extreme consumer since within any population there are likely to be a few individuals with very unusual dietary habits whose exposures are completely unrepresentative of that amongst the vast majority of the population. In the EU, high consumer values are taken as the 95th percentile of consumption.[22]

2.6.2.2 Significance of Exceeding the ADI

Provided intakes for average and high consumers are within the ADI, then it is reasonable to assume that there is very unlikely to be any risk to health among the population. If intakes occasionally exceed the ADI, it is still unlikely that harm will result since the ADI is based on a no-effect level, and not on an effect level, to which a large uncertainty factor has been applied.

If, however, exposure estimates indicate that the ADI may be regularly exceeded by certain groups in the population, then it may be necessary to reduce the permitted concentrations in foods, or reduce the range of foods in which the additive is permitted for use. Since concentrations in foods are determined by what is needed to achieve the desired technological effect, the option to reduce the amounts in foods may not be feasible. In such cases, the food categories in which the additive is allowed to be used may have to be restricted. Reducing permitted use levels is an option when there are several additives within a class (*e.g.* colours, sweeteners, antioxidants) that perform the same function and that may be used in combination with each other.

Even when exposure of high consumers is within the ADI, it may sometimes be necessary to consider those individuals exceeding it to determine whether they represent a discrete population subgroup with particular dietary needs that predispose them to exceed the ADI (*e.g.* exposure of diabetics to a sweetener). Action such as specific targeted advice or modification of particular products may then be necessary.

In cases where the ADI is determined by an effect that can occur following short-term rather than long-term exposure, then particular care may be necessary to ensure the ADI is not exceeded, even on an occasional basis.[41] This would be the case, for example, if the ADI was determined by a developmental effect, since substances affecting embryonic and fetal development can do so after exposure of only a few of days. A more comprehensive discussion of the significance of excursions of intake above the ADI is available.[42]

2.6.3 Re-evaluation of Permitted Additives

EFSA is conducting a re-evaluation of all currently permitted food additives in the EU. A programme, outlining the timelines by which the various classes of additives are to be re-evaluated, starting with colours, has been published.[17] Following public calls for submission of data, EFSA considers not only technical, toxicological and exposure information submitted by manufacturers and other interested parties, but also the published scientific literature. The re-evaluation opinions are published on the EFSA website (http://www. efsa.europa.eu/en/topics/topic/additives.htm). This programme has already resulted in the ADIs for some additives being withdrawn, and consequently their removal from the permitted lists. In other cases previously established ADIs have been confirmed. In some instances, the numerical value of the ADI has been lowered or raised, which may trigger new risk management actions.

2.7 Conclusions

The safety of food additives is an important public-health topic and considerable EU resources are devoted to their risk assessment, risk management and to legal aspects to ensure that processed food is safe for European consumers. Applications for the marketing of new food additives are assessed first by EFSA, which then advises the risk managers in the European Commission and Member States on whether there are any safety concerns. Risk managers then take the decision on whether or not to authorise the additive for food use. The work of EFSA is open and transparent and all its opinions are published and freely available on its website. For many years, permitted food additives were not re-evaluated unless new data, particularly any raising a safety concern, became available. Now, EFSA is systematically re-evaluating all permitted food additives, the majority of which have been on the market for many years. This should provide further confidence in the safety of the food supply.

References

1. European Commission, *Completion of the Internal Market, Community Legislation on Foodstuffs*. Office for Official Publications of the European Communities, Luxembourg, *COM*, 1985, **85**, 603.
2. European Commission. Regulation (EC) No 1333/2008 of the European Parliament and of the Council of 16 December 2008 on food additives. *Off. J. Eur. Union*, 31.12.2008, **L354**, 16–33.
3. European Commission. Regulation (EC) No 1331/2008 of the European Parliament and of the Council of 16 December 2008 establishing a common authorisation procedure for food additives, food enzymes and food flavourings. *Off. J. Eur. Union*, 31.12.08, **L354**, 1–6.
4. European Commission. Council Directive of 21 December 1988 on the approximation of the laws of the Member States concerning food additives authorized for use in foodstuffs intended for human consumption. (87/107/EEC). *Off. J. Eur. Communities*, 11.2.89, **L40**, 27–33.
5. European Commission. Regulation (EC) No 1333/2008 of the European Parliament and of the Council of 16 December 2008 on food additives. *Off. J. Eur. Union*, 31.12.08, **L354**, 16–33.
6. European Commission. Regulation (EC) No 1332/2008 of the European Parliament and of the Council of 16 December 2008 on food enzymes. *Off. J. Eur. Union*, 31.12.08, **L354**, 7–15.
7. European Commission. Regulation (EC) No 1334/2008 of the European Parliament and of the Council of 16 December 2008 on flavourings and certain food ingredients with flavouring properties for use in and on foods and amending Council Regulation (EEC) No 1601/91, Regulations (EC) No 2232/96 and (EC) No 110/2008 and Directive 2000/13/EC. *Off. J. Eur. Union*, 31.12.08, **L354**, 34–50.
8. European Commission. Commission Regulation (EU) No 234/2011 of 10 March 2011 implementing Regulation (EC) No 1331/2008 of the European Parliament and of the Council establishing a common authorisation procedure for food additives, food enzymes and food flavourings. *Off. J. Eur. Union*, 11.3.11, **L64**, 15–24.

9. European Commission. European Parliament and Council Directive 94/35/EC of 30 June 1994 on sweeteners for use in foodstuffs. *Off. J. Eur. Communities*, 10.9.94, **L237**, 3–12.

10. European Commission. European Parliament and Council Directive 94/35/EC of 30 June 1994 on colours for use in foodstuffs. *Off. J. Eur. Communities*, 10.9.94, **L237**, 13–29.

11. European Commission. European Parliament and Council Directive 94/35/EC of 20 February 1995 on food additives other than colours and sweeteners. *Off. J. Eur. Communities*, 18.3.95, **L61**, 1–40.

12. P. J. Wagstaffe, *Food Addit. Contam.*, 1996, **13**, 397.

13. European Commission. Regulation (EC) No 178/2002 of the European Parliament and of the Council of 28 January 2002 laying down the general principles and requirements of food law, establishing the European Food Safety Authority and laying down procedures in matters of food safety. *Off. J. Eur. Communities*, 1.2.2002, **L31**, 1–34.

14. EFSA. Data requirements for the evaluation of food additive applications. Scientific Statement of the Panel on Food Additives and Nutrient Sources added to Food. *The EFSA Journal*, 2009, **1188**, 1–7. Available at: http://www.efsa.europa.eu/en/scdocs/doc/1188.pdf.

15. EFSA. Guidance of the Scientific Panel of Food Contact Material, Enzymes, Flavourings and Processing Aids (CEF) on the Submission of a Dossier on Food Enzymes for Safety Evaluation by the Scientific Panel of Food Contact Material, Enzymes, Flavourings and Processing Aids. *The EFSA Journal*, 2009, **1305**, 1–26. Available at: http://www.efsa. europa.eu/en/efsajournal/doc/1305.pdf.

16. EFSA. Guidance on the data required for the risk assessment of flavourings to be used in or on foods. EFSA Panel on Food Contact Materials, Enzymes, Flavourings and Processing Aids. *The EFSA Journal*, 2010, **8(6)**, 1623. Available at: http://www.efsa.europa. eu/en/efsajournal/doc/1623.pdf.

17. European Commission. Commission Regulation (EU) No 257/2010 of 25 March 2010 setting up a programme for the re-evaluation of approved food additives in accordance with Regulation (EC) No 1333/2008 of the European Parliament and of the Council on food additives. *Off. J. Eur. Union*, 26.03.2010, **L080**, 19–27.

18. European Commission. Commission Directive 95/31/EC of 5 July 1995 laying down specific purity criteria concerning sweeteners for use in foodstuffs. *Off. J. Eur. Communities*, 28.7.95, **L178**, 1.

19. European Commission. Commission Directive 95/45/EC of 26 July 1995 laying down specific purity criteria concerning colours. *Off. J. Eur. Communities*, 22.9.95, **L226**, 1–45.

20. European Commission. Commission Directive 96/77/EC of 2 December 1996 laying down specific purity criteria on food additives other than colours and sweeteners. *Off. J. Eur. Communities*, 30.12.96, **L339**, 1.

21. FAO/WHO. *Principles and Methods for the Risk Assessment of Chemicals in Food. Environmental Health Criteria 240.* A joint publication of the Food

and Agriculture Organization of the United Nations, World Health Organization. 2009, WHO, Geneva. Available at: http://www.who.int/foodsafety/chem/principles/en/index1.html.

22. EFSA. Guidance for submission for food additive evaluations. EFSA Panel on Food Additives and Nutrient Sources to Food (ANS). EFSA Journal 2012;10(7):2760. Available at: http://www.efsa.europa.eu/en/efsajournal/doc/2760.pdf.

23. US FDA. *Toxicological Principles for the Safety Assessment of Food Ingredients. Redbook 2000.* July 2000, updated July 2007. US Food and Drug Administration, Washington DC. Available at: http://www.fda.gov/Food/GuidanceComplianceRegulatoryInformation/GuidanceDocuments/FoodIngredientsandPackaging/Redbook/default.htm.

24. OECD. *Guidelines for the Testing of Chemicals and subsequent revisions.* Organisation for Economic Co-operation and Development, Paris. 2011. Available at: http://www.oecd-ilibrary.org/content/package/chem_guide_pkg-en.

25. European Commission. Commission Directive 84/449/EEC of 25 April 1984, adapting to technical progress for the sixth time, Council Directive 67/548/EEC on the approximation of the laws, regulations and administrative provisions relating to the classification, packaging and labelling of dangerous substances. *Off. J. Eur. Communities*, 19.9.1984, **L251**, 1.

26. European Commission. Commission Directive 87/432/EEC of 3 August 1987, adapting to technical progress for the eighth time, Council Directive 67/548/EEC on the approximation of the laws, regulations and administrative provisions relating to the classification, packaging and labelling of dangerous substances. *Off. J. Eur. Communities*, 21.8.1987, **L239**, 1.

27. European Commission. Commission Directive 92/69/EEC of 31 July 1992, adapting to technical progress for the seventeenth time, Council Directive 67/548/EEC on the approximation of the laws, regulations and administrative provisions relating to the classification, packaging and labelling of dangerous substances. *Off. J. Eur. Communities*, 29.12.1992, **L383**, 1.

28. Scientific Committee on Food. Report of the Scientific Committee for Food on Antioxidants (opinion expressed 11 December 1987). *Reports of the Scientific Committee on Food (Twenty-second Series)*. CEC, Office for Official Publications of the European Communities, Luxembourg, 1989.

29. EFSA. Scientific Opinion on the Re-evaluation of Butylated Hydroxyanisole – BHA (E 320) as a food additive. EFSA Panel on Food Additives and Nutrient Sources added to food (ANS). *The EFSA Journal*, 2011, **9(10)**, 2392. Available at: http://www.efsa.europa.eu/en/efsajournal/pub/2392.htm.

30. Scientific Committee on Food. Report of the Scientific Committee for Food on Sweeteners (opinion expressed on 14 September 1984). *Reports of the Scientific Committee on Food (Sixteenth Series)*. CEC, Office for Official Publications of the European Communities, Luxembourg, 1985.

31. National Research Council. *Carcinogens and Anticarcinogens in the Human Diet*, National Academy Press, Washington DC, 1996, pp. 162–163.

32. EFSA. Scientific Opinion on the Re-evaluation of Erythrosine (E 127) as a Food Additive. EFSA Panel on Food Additives and Nutrient Sources added to Food (ANS). *The EFSA Journal*, 2011, **9(1)**, 1854.
33. EFSA. Assessment of the results of the study by McCann *et al.*, (2007) on the effect of some colours and sodium benzoate on children's behaviour – Scientific Opinion of the Panel on Food Additives, Flavourings, Processing Aids and Food Contact Materials (AFC). *The EFSA Journal*, 2008, **660**, 1. Available at: http://www.efsa.europa.eu/en/efsajournal/doc/660.pdf.
34. EFSA. Scientific Opinion on the use of Natamycin as a Food Additive. *The EFSA Journal*, 2009, **7(12)**, 1412. Available at: http://www.efsa.europa. eu/en/efsajournal/pub/1412.htm.
35. EFSA. Scientific Opinion on the use of Gum Acacia modified with Octenyl Succinic Anhydride (OSA) as a food additive. EFSA Panel on Food Additives and Nutrient Sources added to Food. *The EFSA Journal*, 2010, **8(3)**, 1539. Available at: http://www.efsa.europa.eu/en/efsajournal/ pub/1539.htm.
36. Food Additive Intake: Scientific Assessment and Regulatory Requirements in Europe. Eds: R. Walker, A. Lützow, J. Howlett, and M. Knowles. Workshop organized by ILSI Europe. *Food Addit. Contam.*, 1996, **13**, 383.
37. C. Leclercq, *Food Chem. Toxicol.*, 2007, **45**, 2336.
38. ILSI Europe. *An Evaluation of the Budget Method for Screening Food Additive Intake. Summary Report of an ILSI Europe Food Chemical Intake Task Force*. ILSI Europe, Brussels, Belgium, April 1997.
39. M. J. Gibney and J. Lambe, *Food Addit. Contam.*, 1996, **13**, 405.
40. EFSA, Use of the EFSA Comprehensive European Food Consumption Database in exposure assessment, *The EFSA Journal*, 2011, **2097**, 1.
41. E. D. Rubery, S. M. Barlow and J. H. Steadman., *Food Addit. Contam.*, 1990, **7**, 287.
42. ILSI Europe. *Significance of Excursions of Intake Above the Acceptable Daily Intake (ADI). Report of an ILSI Workshop held in April 1998*. ILSI Europe, Brussels, Belgium, July 1999.

CHAPTER 3

The Development of Food Additive Legislation in Europe

DAVID JUKES

Senior Lecturer in Food Regulation, Department of Food and Nutritional
Sciences, The University of Reading, UK
E-mail: d.j.jukes@reading.ac.uk

3.1 Fundamentals

The origins of the current European Union can be found in the 3 Treaties that
were agreed in the 1950s. Of these, the most important was the Treaty of Rome
establishing the European Economic Community (EEC) – signed in Rome in
March 1957, it came into effect from 1st January 1958. At that time there were
just six member states (France, Germany, Italy, Belgium, Luxembourg and the
Netherlands).

There are a few key points to note:

(1) The original title of the Treaty refers to an "economic" community. The
emphasis is on bringing the economies of the 6 member states closer
together – rather than competing, the countries would cooperate for
mutual benefit. As the EEC developed and transformed, first to the
European Communities (EC) and subsequently the European Union
(EU), broader objectives have been included. However, as will be shown
below, in the development of food legislation the issues of trade and
economics remained paramount until only quite recently.

(2) There was a transfer of power to the various institutions of the EEC –
this loss of sovereignty was to be balanced by benefits that would come

Essential Guide to Food Additives, 4th Edition
Edited by Mike Saltmarsh
© The Royal Society of Chemistry 2013
Published by the Royal Society of Chemistry, www.rsc.org

from working together as a larger unit. To achieve this, the Treaty provided for the adoption of European legal documents that had greater importance than the national legislation. Two key legislative documents were created (and continue to exist to this day). These are:

- **The Regulation**: Under the terms of the Treaty (see Figure 3.1 for the precise wording), Regulations are "directly applicable" in all member states. This means that they are legal documents – they have to be complied with by businesses, they have to be enforced by local inspectors and they have to be used by national courts. Different legal and enforcement systems in the various member states makes the actual process more complex – for example, the European Regulations do not contain enforcement powers or penalties and so these have to be provided by national procedures. The main advantage of the Regulation is that it ensures that the specific legal requirements are common throughout all member states so, for example, businesses know that if a product meets the legal requirements of a European Regulation in one member state then it should (in theory) meet the requirements in all member states.

- **The Directive**: Although intended to have a similar effect to a Regulation, the Directive is only an agreement by member states to alter their own national legislation to implement the agreed legislation. It therefore only achieves its aim when each member state has ensured that its national legislation (or, if appropriate, other administrative provisions) contains the agreed requirements. Directives always include a date when it should be fully effective in all member states – however, this is not always complied with leading to uncertainty for businesses.

Article 288

To exercise the Union's competences, the institutions shall adopt regulations, directives, decisions, recommendations and opinions.

- A regulation shall have general application. It shall be binding in its entirety and directly applicable in all Member States.

- A directive shall be binding, as to the result to be achieved, upon each Member State to which it is addressed, but shall leave to the national authorities the choice of form and methods.

- A decision shall be binding in its entirety. A decision which specifies those to whom it is addressed shall be binding only on them.

- Recommendations and opinions shall have no binding force.

Figure 3.1   Extract from the Treaty on the Functioning of the European Union. Note: This is the current 2010 text. Although the Article number has changed as the Treaty has been updated, the text has remained unchanged from the original 1958 Treaty version.

(3) The Treaty establishing the EEC created a number of "Institutions" to manage the process. In particular, there were four key bodies that have continued to exist although their functions and powers have been modified over the years. These four are:

- **The Commission**: The Commission can be seen as providing direction and management to the development of the Community/Union. The official "Commission" (as mentioned in the Treaty) is the body of Commissioners who are appointed on the basis of nominations by member states. Currently, there is one Commissioner from each member state. However, although nominated on a national basis, as Commissioners they are required to be independent and work to enhance the Community/Union. When these Commissioners meet as the Commission, they can propose legislation for consideration and adoption by the other institutions (see below). In some cases, however, when delegated power has been given to it, the Commission has the authority to adopt legislation on its own. The Commissioners head the "civil service" and run the various departments known as Directorates-General (DG) – these can be considered as similar to structures used within a national government. These DGs are mostly based in Brussels and the term "Commission" is often used to refer to them all collectively. The term "Commission" can therefore be a bit ambiguous.

- **The Council** (originally the "Council of Ministers"): To provide the national focus, member states send Ministers (or suitable equivalents) to attend meetings of the Council. This Institution was the original decision-making body with the greatest authority. In the early days, agreement in the Council had normally to be unanimous (usually not difficult with just 6 member states) but alternative systems of voting had to be developed as the Community/Union expanded. A majority voting system with weightings based on population is now normal. Authority has been gradually lost with the increasing role of the European Parliament.

- **The European Parliament** (originally the "Assembly"): The Assembly was created to provide a wider reflection of the views of the people within the Community. Different procedures were used to nominate these representatives although, since 1979, all member states have held elections. The Assembly itself decided to change its name to the European Parliament and this was subsequently agreed and adopted into the Treaty. Although the idea to be a "parliament" (with all the authority that the name implies) was originally only an aspiration, the powers of the Parliament have steadily increased such that it does have a major role in the adoption of legislation. For example, even where the Commission has delegated power to adopt legislation, there are procedures allowing the European Parliament to override this.

- **The Court of Justice**: As a legal document, the original Treaty and the various later updates (and in fact all subordinate legislation) required a legal system to provide interpretation and to judge any disputes relating to it. The Court of Justice provides for this. Whilst in many cases the judgements of the Court of Justice relate to complex legal arguments, they have the potential to have a significant impact on the overall legal structure and can change the way the Treaty, the Regulations and the Directives are viewed. As will be seen, European legislation on food additives has certainly been influenced by the Court's judgements.

(4) Legislation is adopted following procedures laid down in the Treaty. Originally this was a relatively simple process that, in summary, involved the Commission proposing new legislation and the Council of Ministers reaching a decision on the final document. The Assembly was, however, "consulted" but did not have the power to insist on amendments. Most food legislation now follows a much more detailed process (termed the "ordinary legislative procedure") that involves the Commission, the Council and the European Parliament. A summary of this process is shown in Figure 3.2. However, as already indicated, the Commission is frequently given delegated powers to adopt legislation and this usually applies when matters are considered to be merely technical amendments to agreed controls. This delegated legislation (usually now termed "implementing" measures) has to be adopted following defined procedures. This includes discussions within regulatory committees established by the Commission but including representatives from member states and, under specific circumstances, an opportunity for the Council and the European Parliament to prevent the adoption of the legislation. The Commission's proposals for implementing measures relating to food additives are discussed by the "Standing Committee on the Food Chain and Animal Health" regulatory committee and, in particular, its "Toxicological Safety of the Food Chain" section.

3.2 Stage One – Early Harmonisation Attempts

As indicated already, the original founding Treaty was aiming to establish an "economic" community. Trading would be made easier through the removal of quotas and quantitative restrictions but there would also be agreements to reduce the technical barriers to trade. Quite quickly, national controls on food additives were identified as an area where differences between countries made trade difficult or expensive.

In many countries, the 1950s had been a time when tighter controls on food additives were being agreed. An enhanced scientific understanding of the risks that might arise from the use of additives was leading countries to adopt positive lists of food additives. Earlier controls had frequently been based on "negative lists" that were limited to the banning of substances known to be harmful. The switch to "positive lists" meant that only those additives that had

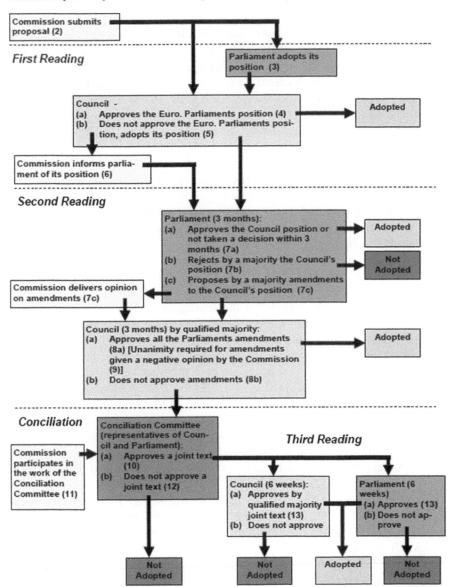

Figure 3.2 The Ordinary Legislative Procedure: A summary based on Article 294 of the Treaty on the Functioning of the European Union.
Note: Numbers refer to subsections of Article 294. Normally time limits are indicated although, with agreement, longer is permitted.

been listed in the legislation could be used, and in most cases, detailed restrictions on their use were included. As these national controls were based on national criteria and priorities using national advisory committees, their development led to different lists and different restrictions.

The concept was to create a single European list with a single agreed set of restrictions – achieving this was, however, to take over 50 years! Most countries had started to adopt their lists of additives based on their functions (colours, preservatives, *etc.*) and this was the procedure used to move to a European system. To help classify each additive, a numbering system was created in which different functional classes were allocated to a specific group of numbers. The first set allocated were food colours that were given numbers starting from 100. Also, to demonstrate that they had a European-wide approval, these numbers were preceded by the letter "E" – the E-number had been born!

There are 3 main components to the legal approval of a food additive:

   (i) a listing of those additives which are permitted;
  (ii) a listing of the foods in which foods these can be used; and
 (iii) limits on the levels at which they can be used in these foods.

With different national dietary habits, getting agreement on these components is easiest with the first and hardest with the last. The first legal document therefore only considered the first issue and attempted to agree a single list of additives that had to be incorporated into national additive controls – it left to individual countries the decision as to which foods and what levels would be acceptable. As the application of the agreed European list would require detailed discussions within each member state, the agreed listing of approved colours was adopted in 1962 in the form of a Directive. It can be noted that for many years the Directive was the preferred method for achieving harmonisation in many different sectors. The colours' Directive was followed quite quickly by a Directive for preservatives in 1964 (with numbers starting from E200) and more slowly by Directives for antioxidants (in 1970 with numbers from E300) and emulsifiers, stabilisers, thickeners and gelling agents (in 1974 with E400 numbers).

It can be noted that the pace of harmonisation was slow even with just 6 member states. In 1973 the membership of the EEC increased to 9 with the United Kingdom, Ireland and Denmark becoming members. This created some transitional complications as additional food additives, not previously permitted by the original 6 member states were in use in the new members. Additional numbers were allocated to these on a temporary basis (without the "E" prefix) pending decisions on whether to grant them full Community acceptance. In describing the political difficulties faced at this time, the Commission stated in 1985:

"The Community lists currently contain some 150 additives. For ten years it has been virtually impossible to obtain unanimous agreement to add to these lists. Preservatives provide a striking example. In 1981, the Commission proposed to authorise or to extend the authorisation of three substances (including natamycin) regarded as perfectly acceptable by the Scientific Committee for Food. So far, after almost four years of discussion, the Council

has been unable to reach agreement on the proposal. At the very best, it could agree on a possibility of national derogations, an absolutely unacceptable solution from the viewpoint of a single market." [†]

3.3 Stage 2 – Creating the Internal Market

It was fully appreciated that the process of removing barriers to trade was taking much longer than anyone had expected – not just with food additives but in many areas of harmonisation. Even with a single list of permitted additives, the different foods in which the additives were permitted and the different levels of use made trade complex and led to additional costs. However, achieving political agreement in an area where consumers were growing increasingly concerned was very difficult for the Ministers who attended the Council meetings where these matters were discussed and had to be agreed.

A possible solution resulted from the landmark "Cassis de Dijon" judgement in the Court of Justice in 1979. This related to the import into Germany of the blackcurrant liqueur, Cassis de Dijon, which has an alcoholic content that fell between permitted bands under German legislation. Briefly, after hearing legal arguments based on the wording of the Treaty, the Court decided that since the drink had been legally made and marketed in France, the German legislation was creating a barrier to trade that was prohibited by the Treaty. Following further similar cases, this led to the realisation that countries should accept products from other countries even when not complying with their own controls (known as "mutual recognition") – it was not necessary to harmonise all legal requirements.

However, the Treaty did provide for countries to maintain national controls when it could be shown to be necessary to protect public health. The control of food additives was clearly a public-health concern and, since it was obvious that consumers' diets varied between member states, it was successfully argued that different national controls on food additives were justified to protect public health.

During the early 1980s the Commission started to focus on dramatically increasing economic integration by the removal of remaining barriers to trade. The resulting programme to create the "Internal Market" was to run from 1985 to 1992. Rather than trying to harmonise all matters, the focus would provide a "new approach" and would concentrate in particular on those areas in which the Treaty provisions allowing a public-health justification to override the "Cassis de Dijon" principle was being used to maintain national provisions and creating trade barriers.

For food additives this resulted in detailed discussions on establishing agreed European controls covering all 3 components mentioned above. Given the political sensitivity of the issue at a time when consumer concern at the use of food additives had increased significantly (under agreed changes to European food labelling, E-numbers had appeared on food labels in the early 1980s), it

---

[†]Commission Document (1985): Communication from the Commission the Council and to the European Parliament – Completion of the internal market: Community legislation on foodstuffs. COM(85)603 final.

was still very difficult to agree a single set of controls. The proposed structure was for a framework Directive to contain general principles and procedures and for separate Directives to specify the permitted substances, foods and levels. There were major arguments on both procedural matters (*e.g.* should delegated authority be given to the Commission for the updating of the list of additives?) and on more detailed technical issues (*e.g.* should countries be allowed to maintain traditional prohibitions on the use of additives in certain foods such as the German controls on "pure beer").

Not only were there arguments between countries but there were also arguments between the Commission, the Council of Ministers and the European Parliament. However, eventually a compromise was achieved that resulted in a framework Directive (Directive 88/107) and three specific directives – one for sweeteners (Directive 94/35), one for colours (Directive 94/36) and one that covered the remaining additives (Directive 95/2). Because of the political sensitivity of the issues, it can be noted that these specific Directives were agreed quite some time after the end of the official timetable for creating the "Internal Market". On the issue of delegated powers, the Commission was forced to accept that it would not be given the authority to update the controls and, as a result, any subsequent amendment to the lists of additives or their uses has had to go through the full procedure involving the Council of Ministers and the European Parliament. On the issue of traditional products, it was agreed to establish a limited list of foods that, under traditional national legal practices, had been subject to restrictions of the use of additives. This was necessary, in particular, to overcome the concerns of Germany, which wanted to ensure that German beer would continue to be produced without the use of additives. This required an amendment (by Directive 94/34) to the framework directive and the list of approved national provisions was adopted in 1996 (and published in early 1997 as Decision 292/97).

Alongside the specific Directives were a set of Directives establishing the purity criteria for each additive – providing in effect a legal specification for each additive. It can be noted that in the case of amendments to the purity criteria, delegated powers were granted to the Commission since these were not considered to be so politically sensitive by member states.

Even though the process was difficult, when adopting maximum limits on food additives for the first time it was necessary to take into account the different diets in the different member states. Where toxicological evidence indicated the need for a limit to be placed on the usage of an additive, compromise agreements were sometimes necessary. It was also necessary to consider how to deal with those additives regarded as safe for which a legal maximum based on toxicology was not appropriate. Most countries agree that it is appropriate to limit the use of additives to cases where it is needed and where it is safe. New terminology was required to cover this situation and the legal term "*quantum satis*" was introduced (for definitions of "*quantum satis*" see Figure 3.3).

Following their adoption it was of course necessary, from time to time, to update the controls. Although the full legislative procedure was involved, amendments were passed that added new additives or which amended the

---

Directive 94/36: Colours

Article 2(7): In the Annexes to this Directive *"quantum satis"* means that no maximum level is specified. However, colouring matters shall be used according to good manufacturing practice at a level not higher than is necessary to achieve the intended purpose and provided that they do not mislead the consumer.

Directive 94/35 (as amended by Directive 96/83): Sweeteners

Article 2(5): In the Annex *"quantum satis"* means that no maximum level is specified. However, sweeteners shall be used in accordance with good manufacturing practice, at a dose level not higher than is necessary to achieve the intended purpose and provided the consumer is not misled.

Directive 95/2: Additives other than colours and sweeteners

Article 2(8): In the Annexes to this Directive, *"quantum satis"* means that no maximum level is specified. However, additives shall be used in accordance with good manufacturing practice, at a level not higher than is necessary to achieve the intended purpose and provided that they do not mislead the consumer.

Regulation 1333/2008: Food Additives

Article 3(2)(h): *"Quantum satis"* shall mean that no maximum numerical level is specified and substances shall be used in accodance with good manufacturing practice, at a level not higher than is necessary to achieve the intended purpose and provided the consumer is not misled.

Figure 3.3 Definitions of "*quantum satis*" in EU additive legislation.
Note: When first adopted Directive 94/35 failed to include the definition. This was corrected by the amendment introduced by Directive 96/83.

permitted levels. A summary of the developing structure during this stage is shown in Figure 3.4. It can be noted that the controls on colours remained unchanged from 1994 until the adoption of the current new controls in 2008.

During this time there was an increased role for scientific advice at the European level. In fact as early as 1974, the Commission had created a "Scientific Committee for Food" (SCF) that could provide the Commission with advice that could then be used as the basis of its proposals. The need to consult the SCF when considering new additives was incorporated into many of the new Directives. The SCF had been very actively consulted during the Internal Market programme but most member states still placed more reliance on their own national advisory bodies.

3.4 Stage Three – Changing Emphasis: from Trade to Public Health

Although the Internal Market system of additive controls was a major advance, the motivation had been the removal of barriers to trade. During the late 1980s and the early 1990s, consumer confidence in both national and European food

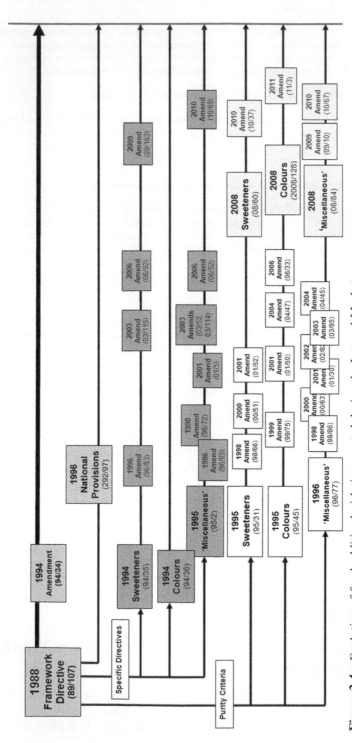

Figure 3.4   Evolution of food additive legislation created during the Internal Market programme.
Note: In 2008 the 3 Directives providing the purity criteria were subject to a process of legal consolidation so as to provide an updated document incorporating the amendments which had been adopted following its first publication.

safety controls dropped dramatically. The initial cause was largely the growing concern over the disease Bovine Spongiform Encephalopathy (BSE) and the subsequent acceptance by politicians (both in the UK, in 1996, and elsewhere) that the public had been poorly informed and inadequately protected. Other crises also occurred including one involving the chemical dioxin that was found to have contaminated a wide range of foods in Belgium (and neighbouring countries) in early 1999.

For our purposes it is not necessary to give all the details but during the late 1990s there was a major shift with the recognition that European food legislation should be shaped with public health at the core. One of the difficulties during the BSE crisis had been that different national governments were still basing their decisions upon national scientific advisory bodies and that although the European advisory committees (such as the SCF) were making recommendations, these were not always accepted by member states. The main result of this dramatic shift was that Regulation 178/2002 was adopted that established a new EU-wide framework for food law and that created the European Food Safety Authority (EFSA) as an independent scientific advisory body.

Another change that came about at this time was the move from adopting Directives to adopting Regulations. As mentioned at the start of this chapter, Regulations are "directly applicable", whereas Directives need to be adopted into national legislation. The EEC had mutated into, first, the EC and then the European Union (EU). At the same time, with the growth in members from the original 6 to 9, then to 10, 12, 15, 25 and now 27, the task for the Commission of ensuring that member states have implemented the agreed controls became much harder. It was also almost impossible for food businesses to know whether national legislation had been fully updated to meet the Directives (some countries were known to be much slower than others in the process of transposition!). Therefore, from the late 1990s it became normal for new legislation (including that for food) to be proposed as a Regulation.

In January 2000 the Commission published its plans in a "White Paper on Food Safety" – a policy statement outlining plans for a major new programme of legislation. With respect to food additives, the Commission stated the following:

"The provisions relating to food additives and flavourings need to be amended in several respects. Firstly, implementing powers should be conferred on the Commission to maintain the Community lists of authorised additives and the status of enzymes should be clarified. Secondly, the Community lists of colouring matters, sweeteners and other additives need to be updated. Thirdly, the purity criteria for sweeteners, colours and other additives have to [be] amended and appropriate purity criteria for food additives made from novel sources have to be laid down. The Commission will further publish a report on the intake of food additives."[‡]

[‡]Commission Document (2000): White Paper on Food Safety. COM(1999)719 final.

Based on this programme, the new framework for food law was provided by Regulation 178/2002 and this framework has been used as the basis for updating much existing food law. A major change was the adoption of the new "hygiene package" in 2004 but, as indicated in the above quote, it had also been noted that the food additive framework needed additional work. With the early work on additives, the approach had been largely driven by the need to remove barriers to trade. It had been a fragmented process driven by a desire to use food legislation to fix the trade problems. The new focus meant that there was an opportunity to start creating an overall structure for food legislation – a structure with consumer protection as the priority.

The updating of the legislation suggested by the quotation given above proved to be much more radical. Although incorporating many of the previous elements, the opportunity was taken to devise a fundamentally new structure. The disjointed approach previously adopted had, for example, led to the creation of separate procedures and controls for flavourings. There were also gaps – controls on enzymes used in food processing had not been developed (except where their use meant that they were clearly classified as food additives under the definition of a "food additive" contained in the framework directive). Detailed discussion therefore led to the creation of a new structure – one that was both more protective of the public but also more accessible for the users.

Although detailed information on the new Regulations is provided in the following Chapters, it might be useful to give an overview here. The new framework Regulation (Regulation 1331/2008) is established by a single regulation setting out the principles of the controls that incorporate scientific risk assessment provided by EFSA. The risk assessment then forms the basis of the risk-management decisions taken by the Commission, the Council and the European Parliament that result in the legislation. In contrast to the controls agreed during the Internal Market programme, delegated authority has now been given to the Commission to amend the detailed technical requirements.

This framework is then used as the basis for the detailed controls. In fact it is used for three different specific Regulations: food additives (Regulation 1333/2008), food flavourings (Regulation 1334/2008) and food enzymes (Regulation 1332/2008). It is also likely that additional sectors may be added – it had been proposed that updated controls on novel foods would be a fourth sector subject to Regulation 1331/2008 but agreement of this proposal has been delayed. Currently, the whole area is known as "food improvement agents" reflecting its wider coverage than just "food additives" although this issue arises more from the EU's legal definition of "food additive" that specifically excludes flavourings. As a result a broader term has to be used.

Although when first published, Regulation 1333/2008 on food additives contained Annexes for the listing of the details of the permitted additives, the foods in which they could be used and the maximum levels of use, these Annexes were initially empty. Detailed work continued on the content of the Annexes and these were subsequently published in amending Regulations (Regulations 1129/2011 and 1130/2011) and these have already been subject to further minor updates. The main point to make here is that the structure of these Annexes has totally changed from the previous controls. Instead of being based on a functional classification of the additives, the Annex is now based on a classification of foods. Thus, once a food manufacturer has identified the correct classification for their product, they can easily determine the complete list of additives that are permitted and at what levels. It is therefore now based on the needs of the user rather than what is simplest for the legislator. This does of course also make it easier for the enforcement officers when checking to see whether the legislation is being applied correctly!

At the time of writing, the EU is still in transition from the controls based on the Internal Market Directives to the new system based on Regulations. However, it does at last reflect a unified system of controls that should be capable of meeting the needs of the EU for many years. It is the nature of European developments that political compromises are often needed to make progress. The willingness to make those compromises is not always sufficiently strong. Whilst this can be frustrating for those subject to the legislation, that is the nature of the society in which we live.

CHAPTER 4

European Legislative Framework Controlling the Use of Food Additives

ANNIE-LAURE ROBIN*[a] AND DEVINA SANKHLA[b]

[a] International Regulatory Manager at Leatherhead Food Research, UK;
[b] Senior Regulatory Advisor at Leatherhead Food Research, UK
*E-mail: arobin@leatherheadfood.com

4.1 Introduction

Within the European food legislative framework, the approval of a substance as a food additive is intended to both ensure the protection of public health regarding the use of food additives in foodstuffs and inform consumers about their presence in food products. The use of food additives is strictly regulated in the European Union. The legislation covers their definitions, exemptions, approvals, category names, additive names, synonyms and E numbers, purity criteria, conditions of use and labelling. In addition, trends towards more "fresh" foods and the growth in market share of chilled foods, together with changes in legislation following the completion of the European legislative harmonisation exercise in 2008, have all had an impact on the use of food additives in Europe.

This chapter therefore provides an overview of the definitions and controls of the use of food additives under the current and future European food legislative framework, whilst bearing in mind that these substances will always be essential to food preparation, quality and preservation.

Essential Guide to Food Additives, 4th Edition
Edited by Mike Saltmarsh
© The Royal Society of Chemistry 2013
Published by the Royal Society of Chemistry, www.rsc.org

4.2 EU Food Law Framework

The European Regulatory framework is made up of a variety of legislation controlling foodstuffs by specific topics. These have been organised in a structured way specific to the European Union. Above all, Regulation (EC) No. 178/2002 (as amended) on general food law[1] lays the foundations of all food legislation including legislation for food additives. It provides a basis for the assurance of a high level of consumer protection. Figure 4.1 provides a simplified overview of the structure of European food law by main topics.

The legislation for food additives is quite complex and has changed significantly over the past few years. As summarised in Figure 4.2, in 1989, a

Figure 4.1 Simplified overview of the structure of European food law by main topics.

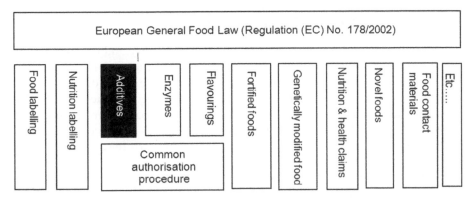

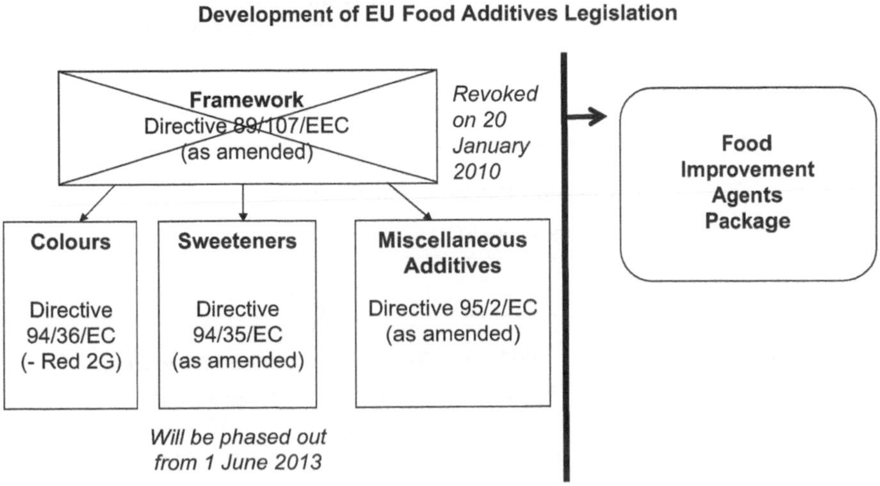

Figure 4.2 Simplified overview of the development of European food legislation on food additives.

European food additives framework was published in Council Directive 89/107/EEC.[2] This Directive set out criteria for the authorisation of food additives in the European Union, as well as definitions, general criteria for the use of food additives, and labelling information. The Directive had been amended twice and was complemented by three specific Directives namely 94/35/EC on sweeteners,[3] 94/36/EC on colours,[4] and 95/2/EC on miscellaneous food additives (other than colours and sweeteners).[5] These Directives had been amended several times for sweeteners and miscellaneous food additives, and one colour 'Red 2G' had been suspended due to safety reasons.[32] These Directives were also implemented into national law by each European Member State.

4.2.1   EU Food Improvement Agents Package

In 2006, the European Commission proposed to simplify and consolidate the legislation controlling food additives, as well as the legislation for food enzymes and food flavourings. These proposed Regulations were later adopted in 2008 under the so-called European Food Improvement Agents Package (FIAP). This package replaced the legislation previously laid down in the framework Directive 89/107/EEC for food additives from 20 January 2010, and phases out the European Directives for colours, sweeteners and miscellaneous food additives from 1 June 2013.

As shown in Figure 4.3, the FIAP comprises the following four Regulations:

- Regulation (EC) No. 1331/2008 of 16 December 2008 establishing a common authorisation procedure for food additives, food enzymes, and food flavourings;[6]
- Regulation (EC) No. 1332/2008 of 16 December 2008 on food enzymes;[7]
- Regulation (EC) No. 1333/2008 of 16 December 2008 on food additives;[8]

EU Food Improvement Agents Package

Common Authorisation Procedure
Regulation (EC) No. 1331/2008

Food Enzymes	Food Additives	Food Flavourings
Regulation (EC) No. 1332/2008	Regulation (EC) No. 1333/2008	Regulation (EC) No. 1334/2008

Figure 4.3   European Food Improvement Agents Package (FIAP).

- Regulation (EC) No. 1334/2008 of 16 December 2008 on flavourings and certain food ingredients with flavouring properties for use in and on foods.[9]

4.2.2 EU Regulation (EC) No. 1333/2008 on Food Additives

Regulation (EC) No. 1333/2008 (as amended) on food additives[8] applies from 20 January 2010. This Regulation simplifies the old legal framework on food additives by grouping food colours, food sweeteners, and other food additives under the same piece of legislation. These have been controlled since 1994 under three different European Directives,[3-5] the implementation of which differed slightly between European Member States. The advantage of this Regulation is that it will apply directly in all European Member States.

The Regulation (EC) No. 1333/2008 (as amended) contains the following five annexes:

- Annex I lists the 26 EU functional classes of food additives;
- Annex II contains the Union list of food additives approved for use in foods and their conditions of use;[10]
- Annex III contains the Union list of food additives including carriers approved for use in food additives, food enzymes, food flavourings, nutrients and their conditions of use;[11]
- Annex IV lists traditional foods for which certain Member States may continue to prohibit the use of certain categories of food additives (Table 4.1);
- Annex V lists the food colours for which the labelling of foods must include additional information.

4.2.3 EU Purity Criteria for Food Additives

All authorised food additives must fulfil purity criteria that are currently detailed in three Directives: Directive 2008/60/EC (as amended) for sweeteners,[12] Directive 2008/128/EC (as amended) for colours,[13] and Directive 2008/84/EC (as amended) for miscellaneous food additives (other than colours and sweeteners).[14]

Following the establishment of the Union lists of food additives into Annexes II and III to Regulation (EC) No. 1333/2008 (as amended), the purity criteria for existing food additives have undergone a review and consolidation by the Commission, which created a new Regulation [Regulation (EU) No. 231/2012], with specifications for food additives, which came into force on the 11 April 2012.[15]

This Regulation will repeal the previous Directives for colours, sweeteners and miscellaneous food additives from 1 December 2012, from which point the new specifications will apply to existing food additives. However, the specifications apply now for new food additives such as steviol glycosides (E960).[15]

Table 4.1   Traditional foods for which certain Member States may continue to prohibit the use of certain categories of food additives.

Member State	Foods	Categories of additives that may continue to be banned
Germany	Traditional German beer (Bier nach deutschem Reinheitsgebot gebraut)	All except propellant gases
France	Traditional French bread	All
France	Traditional French preserved truffles	All
France	Traditional French preserved snails	All
France	Traditional French goose and duck preserves (confit)	All
Austria	Traditional Austrian "Bergkäse"	All except preservatives
Finland	Traditional Finnish "Mämmi"	All except preservatives
Sweden Finland	Traditional Swedish and Finnish fruit syrups	Colours
Denmark	Traditional Danish "Kødboller"	Preservatives and colours
Denmark	Traditional Danish "Leverpostej"	Preservatives (other than sorbic acid) and colours
Spain	Traditional Spanish "Lomo embuchado"	All except preservatives and antioxidants
Italy	Traditional Italian "Mortadella"	All except preservatives, antioxidants, pH-adjusting agents, flavour enhancers, stabilisers and packaging gas
Italy	Traditional Italian "Cotechino e zampone"	All except preservatives, antioxidants, pH-adjusting agents, flavour enhancers, stabilisers and packaging gas

4.3   EU Food Categories for the Use of Food Additives

Regulation (EC) No. 1333/2008 (as amended) introduces a new way of presenting the conditions of use of food additives compared to the EU Directives.[3-5] In these EU Directives, food additives were presented against their specific conditions of use, which meant that it was time consuming to assess and summarise all food additives that could be used in a given foodstuff. The Regulation, however, seems more user-friendly by presenting this information the other way round, *i.e.* each food product category is presented with the list of food additives that it may contain. The nineteen different EU food product categories and their subcategories are listed in Annex II, Part D of Regulation (EC) No. 1333/2008 (as amended), as follows:

0. All categories of foods;
1. Dairy products and analogues;
2. Fats and oils and fat in oil emulsions;
3. Edible ices;
4. Fruit and vegetables;
5. Confectionery;

6. Cereals and cereal products;
7. Bakery wares;
8. Meat;
9. Fish and fisheries products;
10. Eggs and egg products;
11. Sugars, syrups, honey and table-top sweeteners;
12. Salts, spices, soups, sauces, salads and protein products;
13. Foods intended for particular nutritional uses as defined by Directive 2009/39/EC;
14. Beverages;
15. Ready-to-eat savouries and snacks;
16. Desserts excluding products covered in categories 1, 3 and 4;
17. Food supplements as defined in Directive 2002/46/EC of the European Parliament and of the Council excluding food supplements for infants and young children;
18. Processed foods not covered by categories 1 to 17, excluding foods for infants and young children.

These categories also contain subcategories that are not detailed above but can be viewed in Part D of Annex II to Regulation (EC) No. 1333/2008 (as amended).[8,10]

4.4 EU Food Additives

4.4.1 Definition

The EU legal definition of a food additive is laid down in Article 3 (2)(a) of Regulation (EC) No. 1333/2008 (as amended) on food additives[8] as follows:

> *"any substance not normally consumed as a food in itself and not normally used as a characteristic ingredient of food, whether or not it has nutritive value, the intentional addition of which to food for a technological purpose in the manufacture, processing, preparation, treatment, packaging, transport or storage of such food results, or may be reasonably expected to result, in it or its by-products becoming directly or indirectly a component of such foods."*

Additionally, a food additive must comply with specific criteria regarding its safety,[12–15] technological needs and conditions of use in foodstuffs, and its benefits to consumers must be justified.

4.4.2 What is not Considered a Food Additive in the EU?

Regulation (EC) No. 1333/2008 (as amended) does not consider the following as food additives:

- monosaccharides, disaccharides or oligosaccharides and foods containing these substances used for their sweetening properties;

- foods, whether dried or in concentrated form, including flavourings incorporated during the manufacturing of compound foods, because of their aromatic, sapid or nutritive properties together with a secondary colouring effect;
- substances used in covering or coating materials, which do not form part of foods and are not intended to be consumed together with those foods;
- products containing pectin and derived from dried apple pomace or peel of citrus fruits or quinces, or from a mixture of them, by the action of dilute acid followed by partial neutralisation with sodium or potassium salts (liquid pectin);
- chewing gum bases;
- white or yellow dextrin, roasted or dextrinated starch, starch modified by acid or alkali treatment, bleached starch, physically modified starch and starch treated by amylolitic enzymes;
- ammonium chloride;
- blood plasma, edible gelatin, protein hydrolysates and their salts, milk protein and gluten;
- amino acids and their salts other than glutamic acid, glycine, cysteine and cystine and their salts having no technological function;
- caseinates and casein;
- inulin.

Additionally, it is important to note that the following are also not considered food additives in the EU:

- processing aids;
- substances used for the protection of plants and plant products in accordance with community rules relating to plant health (e.g.: pesticides, herbicides, insecticides);
- substances added to foods as nutrients (e.g.: minerals or vitamins);
- substances used for the treatment of water for human consumption falling within the scope of Council Directive 98/83/EC on drinking water quality;[16]
- flavourings as they are regulated under Regulation (EC) No. 1334/2008 (as amended) on flavourings and certain food ingredients with flavouring properties;
- food enzymes as they are controlled under Regulation (EC) No. 1332/2008 on food enzymes;
- extraction solvents that are subject to specific legislation on both their use and residual levels, under Directive 2009/32/EC (as amended).[17]

4.4.3   EU Functional Classes of Food Additives and their Definitions

These are the 26 European food additive functional classes as listed in Annex I of Regulation (EC) No. 1333/2008 (as amended):[8]

1. Sweeteners;
2. Colours;
3. Preservatives;

 4. Antioxidants;
 5. Carriers;
 6. Acids;
 7. Acidity regulators;
 8. Anticaking agents;
 9. Antifoaming agents;
10. Bulking agents;
11. Emulsifiers;
12. Emulsifying salts;
13. Firming agents;
14. Flavour enhancers;
15. Foaming agents;
16. Gelling agents;
17. Glazing agents (including lubricants);
18. Humectants;
19. Modified starches;
20. Packaging gases;
21. Propellants;
22. Raising agents;
23. Sequestrants;
24. Stabilisers;
25. Thickeners;
26. Flour treatment agents.

These functional classes are defined as follows:

 1. "sweeteners" are substances used to impart a sweet taste to foods or in table-top sweeteners;
 2. "colours" are substances that add or restore colour in a food, and include natural constituents of foods and natural sources that are normally not consumed as foods as such and not normally used as characteristic ingredients of food. Preparations obtained from foods and other edible natural source materials obtained by physical and/or chemical extraction resulting in a selective extraction of the pigments relative to the nutritive or aromatic constituents are colours within the meaning of this Regulation;
 3. "preservatives" are substances that prolong the shelflife of foods by protecting them against deterioration caused by micro-organisms and/or that protect against growth of pathogenic micro-organisms;
 4. "antioxidants" are substances that prolong the shelflife of foods by protecting them against deterioration caused by oxidation, such as fat rancidity and colour changes;
 5. "carriers" are substances used to dissolve, dilute, disperse or otherwise physically modify a food additive or a flavouring, food enzyme, nutrient and/or other substance added for nutritional or physiological purposes to a food without altering its function (and without exerting any technological effect themselves) in order to facilitate its handling, application or use;

6. "acids" are substances that increase the acidity of a foodstuff and/or impart a sour taste to it;
7. "acidity regulators" are substances that alter or control the acidity or alkalinity of a foodstuff;
8. "anticaking agents" are substances that reduce the tendency of individual particles of a foodstuff to adhere to one another;
9. "antifoaming agents" are substances that prevent or reduce foaming;
10. "bulking agents" are substances that contribute to the volume of a foodstuff without contributing significantly to its available energy value;
11. "emulsifiers" are substances that make it possible to form or maintain a homogenous mixture of two or more immiscible phases such as oil and water in a foodstuff;
12. "emulsifying salts" are substances that convert proteins contained in cheese into a dispersed form and thereby bring about homogenous distribution of fat and other components;
13. "firming agents" are substances that make or keep tissues of fruit or vegetables firm or crisp, or interact with gelling agents to produce or strengthen a gel;
14. "flavour enhancers" are substances that enhance the existing taste and/or odour of a foodstuff;
15. "foaming agents" are substances that make it possible to form a homogenous dispersion of a gaseous phase in a liquid or solid foodstuff;
16. "gelling agents" are substances that give a foodstuff texture through formation of a gel;
17. "glazing agents" (including lubricants) are substances that, when applied to the external surface of a foodstuff, impart a shiny appearance or provide a protective coating;
18. "humectants" are substances that prevent foods from drying out by counteracting the effect of an atmosphere having a low degree of humidity, or promote the dissolution of a powder in an aqueous medium;
19. "modified starches" are substances obtained by one or more chemical treatments of edible starches, which may have undergone a physical or enzymatic treatment, and may be acid or alkali thinned or bleached;
20. "packaging gases" are gases other than air, introduced into a container before, during or after the placing of a foodstuff in that container;
21. "propellants" are gases other than air that expel a foodstuff from a container;
22. "raising agents" are substances or combinations of substances that liberate gas and thereby increase the volume of a dough or a batter;
23. "sequestrants" are substances that form chemical complexes with metallic ions;
24. "stabilisers" are substances that make it possible to maintain the physicochemical state of a foodstuff; stabilisers include substances which enable

the maintenance of a homogenous dispersion of two or more immiscible substances in a foodstuff, substances that stabilise, retain or intensify an existing colour of a foodstuff and substances that increase the binding capacity of the food, including the formation of crosslinks between proteins enabling the binding of food pieces into reconstituted food;

25. "thickeners" are substances which increase the viscosity of a foodstuff;

26. "flour treatment agents" are substances, other than emulsifiers, which are added to flour or dough to improve its baking quality.

4.4.4 Carried-Over and Reverse Carried-Over Food Additives

In the EU, a food additive is permitted to be present in a final product made of more than one ingredient (also called a compound foodstuff) even if this food additive is not allowed in this final product, as long as the food additive is permitted in one of the ingredients of the compound food. This is called the carry-over principle. This principle does not apply to: infant formula, follow on formula, processed cereal-based foods, baby foods and dietary foods for special medical purposes intended for infants and young children, except where specially provided for. In addition, Annex II, Part A of Regulation (EC) No. 1333/2008 (as amended)[8,10] lists the exemptions from the use of the carry-over principle for foods where additives are not allowed by the principle, such as honey or unprocessed foods (Table 4.2) and foods where colours are not allowed by the principle, such as in bread and similar products (Table 4.3).

In the EU, there is also the concept of reverse carry-over principle for food additives. This is when an additive authorised in a compound foodstuff can be brought into the foodstuff via an ingredient even if it is not allowed to be used in this ingredient, as long as the ingredient is only used for the purpose of producing the final product in question.

Table 4.2 Foods in which the presence of an additive is not permitted as per the carry-over principle.

Unprocessed foods
Honey
Nonemulsified oils and fats of animal or vegetable origin
Butter
Unflavoured pasteurised and sterilised (including UHT) milk and unflavoured plain pasteurised cream (excluding reduced fat cream)
Unflavoured fermented milk products, not heat treated after fermentation
Unflavoured buttermilk (excluding sterilised buttermilk)
Natural mineral water
Coffee (excluding flavoured instant coffee) and coffee extracts
Unflavoured leaf tea
Sugars
Dry pasta, excluding gluten-free and/or pasta intended for hypoproteic diets

Table 4.3   Foods in which the presence of a food colour is not permitted by virtue of the carry-over principle.

Unprocessed foods

All bottled or packed waters
Milk, full fat, semiskimmed and skimmed milk, pasteurised or sterilised (including UHT sterilisation) (unflavoured)
Chocolate milk
Fermented milk (unflavoured)
Preserved milks (unflavoured)
Buttermilk (unflavoured)
Cream and cream powder (unflavoured)
Oils and fats of animal or vegetable origin
Ripened and unripened cheese (unflavoured)
Butter from sheep and goats' milk
Eggs and egg products
Flour and other milled products and starches
Bread and similar products
Pasta and gnocchi
Sugar including all mono- and disaccharides
Tomato paste and canned and bottled tomatoes
Tomato-based sauces
Fruit juice and fruit nectar and vegetable juice and vegetable nectars
Fruit, vegetables (including potatoes) and mushrooms — canned, bottled or dried; processed fruit, vegetables (including potatoes) and mushrooms
Extra jam, extra jelly, and chestnut purée; crème de pruneaux
Fish, molluscs and crustaceans, meat, poultry and game as well as their preparations, but not including prepared meals containing these ingredients
Cocoa products and chocolate components in chocolate products
Roasted coffee, tea, herbal and fruit infusions, chicory; extracts of tea and herbal and fruit infusions and of chicory; tea, herbal and fruit infusions and cereal preparations for infusions, as well as mixes and instant mixes of these products
Salt, salt substitutes, spices and mixtures of spices
Wine and other products
Spirit drinks, spirits (preceded by the name of the fruit) obtained by maceration and distillation and London gin, Sambuca, Maraschino, Marrasquino or Maraskino and Mistrà
Sangria, Clarea and Zurra
Wine vinegar
Foods for infants and young children including foods for special medical purposes for infants and young children
Honey
Malt and malt products

4.5   Difference Between a Food Additive and a Processing Aid

Some substances can be used either as food additives or as processing aids in the EU.

A substance not normally consumed as a food by itself, used intentionally in the processing of food, only remaining as a residue in the final food, and without any technological effect in the final product is a processing aid.[8]

Therefore, the difference between food additives and processing aids is quite often misunderstood as they are both used for their specific technological functions. In order to clarify this misunderstanding, here are the key differences:

- processing aids are used only during treatment or processing of a food product, and;
- processing aids may result in the nonintentional but unavoidable presence of residues or derivatives, which might not have any technological effects on the finished product.

Therefore, to determine whether a chemical substance is a food additive or a processing aid, one needs to determine whether it continues to function in the final product. For example, residues of a mould-release agent for a confectionery product are unlikely to have a technological effect in the final product, whereas an antioxidant or a preservative added to protect ingredients during processing could still exert a technological function if carried over into the finished product. The former would be considered a processing aid, whilst the latter would be considered a food additive in the EU.

4.6 How to Determine Whether a Substance Is a Food Additive or a Nutrient?

Figure 4.4 summarises the way to classify a substance as a food additive or a nutrient. It is very important to look at the purpose of the substance in the final foodstuff in line with the food additive definition provided in Regulation (EC) No. 1333/2008 (as amended).[8] The main criterion that will determine if a substance is a food additive or a nutrient is in relation to its technological purpose in the final foodstuff. If it has not got any technological purpose but has instead a nutritional purpose, such as isomaltulose or trehalose produced with a new enzymatic process, this substance would be considered a nutrient. The next question is whether this nutrient has a significant history of consumption in the EU before 15 May 1997. If not, it would be considered a novel food ingredient in the EU and must be submitted to a premarket approval assessment under the Novel Food Regulation (EC) No. 258/97 (as amended).[22]

4.7 European Approval Process for Food Additives

Figure 4.5 summarises the European approval procedure for food additives. It is clear from this diagram that there are time limits laid down in the legislation for this procedure that are intended to expedite the process. This was not the

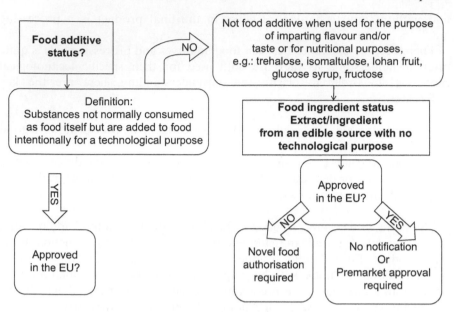

Figure 4.4   Summary diagram on how to classify a substance as a food additive or a nutrient.

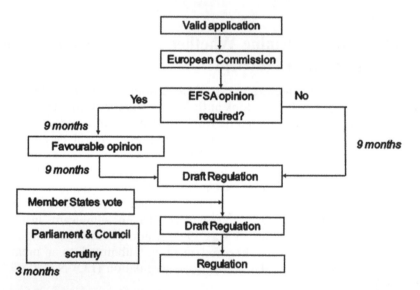

Figure 4.5   European approval procedure for food additives.

case in the past. This new procedure is intended to be more effective and transparent.

Regarding the submission of applications for the approval of new food additives, Regulation (EC) No. 1331/2008[6] covers the common authorisation procedure for food additives. It also lays down procedural arrangements for updating lists of substances, the marketing of which is authorised in the European Union due to Regulation (EC) No. 1333/2008 on food additives.[8]

The procedure for approving a new food additive may be started at the initiative of the Commission or on receipt of an application from an EU Member State or an interested party to the Commission. The interested party who submits the application may represent more than one other interested party. Valid applications are sent by a Member State or interested party directly to the Commission, who acknowledges receipt within fourteen working days. The Commission is responsible for checking their validity.

The European Food Safety Authority (EFSA) will give its opinion on each submission within nine months of receipt. This may be extended if further information from the applicant is required. This is referred to as the "stop the clock" procedure. The Commission may also require further information for risk-management purposes within a stated period. If EFSA issues a positive opinion on the food additive, the Commission then has a further nine months to submit a draft Regulation to include a new substance in the Union list.

Applicants can protect the confidentiality of certain aspects of their application; however, information relating to safety cannot be confidential. In addition, the name and address of the applicant, a clear description of the substance and justification for use as well as analytical methods, where applicable, remain confidential.

Between them, Regulations (EC) No. 1331/2008 and 1333/2008 introduce changes to the Regulation of food additives in the European Union and have positive implications for companies intending to submit food additive dossiers. Under the previous set of food additive Directives,[2–5] the authorisation of a new food additive was subject to the codecision procedure involving votes by the European Parliament and the Council of the European Union before the Commission and Member States could make a decision. This procedure was very lengthy and it was taking up to three years for a food additive to be authorised. The new approval procedure allows a more efficient and simplified procedure for authorisation of food additives by a Comitology procedure involving only votes by Member States in a Committee meeting chaired by the Commission.

It is also worth noting that the European Parliament and Council have a three-month "right of scrutiny" on a decision made by the European Commission and Member States. In most cases, no objections are raised. However, this is not always the case as was seen in 2011 with the "meat glue" food additive (an enzyme preparation based on thrombin and fibrinogen derived from cattle and/or pigs proposed as a food additive for reconstituting food). While EFSA did not identify any safety concerns, and the Commission included this in a draft (SCF) Directive authorising its use, which Member States approved, the Parliament was of the view that it was misleading to consumers and it did not pass this final authorisation step.

In March 2011, the European Commission adopted Commission Regulation (EU) No. 234/2011 which implements Regulation (EC) No. 1331/2008, and sets out requirements for the presentation and content of applications for food additives, food enzymes and food flavourings.[19]

This Regulation was published following guidance from EFSA, and also describes the procedures for validity checks of applications by the European Commission and what must be included in the opinions from EFSA.

The Regulation entered into force on 11 September 2011. Therefore applications for new food additives or extensions of use from 11 September 2011 will be subject to the procedures and deadlines outlined in Commission Regulation (EU) No. 234/2011.

In parallel, the European Commission has developed informal guidance covering some of the practical elements of making an application and this is available on the link below: http://ec.europa.eu/food/food/fAEF/authorisation_ application_en.htm.[20]

4.7.1 European Union Lists of Approved Food Additives

Food additives were authorised under Directives 89/107/EEC (framework Directive) and 94/36/EC (colours), 94/35/EC (sweeteners), and 95/2/EC (miscellaneous food additives) and their amendments. These food additives have been reviewed as required by Regulation (EC) No. 1333/2008 (as amended) on food additives.

Annexes II and III of Regulation (EC) No. 1333/2008 (as amended) now contains Union lists of approved food additives in foods, the conditions of use of the additives in foods, and a list of food additives permitted for use in food additives, food flavourings, food enzymes and nutrient carriers.

When a food additive is contained in the Union list and there is a significant change in the production method or in the starting materials used, or there is a change in particle size, for example, through the use of nanotechnology, such change will require re-authorisation. Also, any genetically modified (GM) food additives will need to be approved under Regulation (EC) No. 1829/2003 (as amended) on GM foods,[21] as well as under this Regulation, to be included in the Union list.

The review of food additives for entry into Annexes II and III was completed on 20 January 2011. The Union list of food additives under Annex II[10] is fully applicable from 1 June 2013 and has included some amendments, such as the inclusion of the conditions of use for steviol glycosides (E 960)[23], the colours quinoline yellow (E 104), sunset yellow FCF (E 110), ponceau 4R or cochineal red A (E 124),[24] aluminium-containing additives,[25] polydextrose (E 1200)[26] and lysozyme (E 1105)[27] in beer, and glycerol esters of wood rosins (E 445) for printing on hard-coated confectionery products.[28]

Annex III contains the Union list of food additives including carriers approved for use in food additives, food enzymes, food flavourings, nutrients and their conditions of use[11] and applies from 2 December 2011, with transitional periods.

4.7.2 EU Food Additives Database

The European Commission Food Additives Database[31] can be used as a tool to inform manufacturers of the food additives approved for use in food in the EU and their conditions of use. The content of this database is based on the Union list provided in Annex II of Regulation (EC) No. 1333/2008 (as amended) and can be accessed on the following link: https://webgate. ec.europa.eu/sanco_foods/main/?event=display

Note that this database has no legal value. The Commission declines all responsibility or liability whatsoever for errors or deficiencies in this database. The official authorisations of food additives are published in the Official Journal of the European Union.[8]

4.8 EU Labelling Rules on Food Additives

4.8.1 Current and Future EU Labelling Legislation

The current European food labelling rules are laid down in Directive 2000/13/EC (as amended).[29] This states that food additives must be declared in the ingredient list by declaring their category names, followed by their E number or legal names, for example:

Ingredients: X,Y, emulsifier: mono-diglycerides of fatty acids, Z.
Or
Ingredients: X, Y, emulsifier: E471, Z.

It must be noted that if an additive belongs to more than one category it has to be declared by the principal function it serves.

Currently, there are 22 listed category names in Directive 2000/13/EC (as amended):[29]

1. Colour;
2. Preservative;
3. Antioxidant;
4. Emulsifier;
5. Thickener;
6. Gelling Agent;
7. Stabiliser;
8. Flavour Enhancer;
9. Acid;
10. Acidity Regulator;
11. Anticaking Agent;
12. Modified Starch;
13. Sweetener;
14. Raising Agent;
15. Antiforming Agent;
16. Glazing Agent;
17. Emulsifying Salts;
18. Flour Treatment Agent;
19. Firming Agent;
20. Humectant;

21. Bulking Agent;
22. Propellant Gas.

Under the new Food Information Regulation for labelling under Regulation (EU) No. 1169/2011,[30] which came into force on 12 December 2011, two new category names, sequestrant and foaming agent, have been added for labelling purposes.

Food businesses have a three-year transitional period to comply with general labelling particulars, until 13 December 2014. Foods placed on the market or labelled prior to 13 December 2014, which do not comply with the requirements of this Regulation, may be marketed until stocks of the foods are exhausted.

Carried-over food additives and processing aids are exempt from being declared in the ingredient lists of food products, except when they are allergenic.

4.8.2   Allergen Declaration

The list of EU allergens is provided in Table 4.4. Under the current European labelling rules on allergen labelling as provided for by Directive 2000/13/EC (as amended),[29] food additives that are derived from a listed allergen must be declared by their specific additive name not E number, for example:

Ingredients: X,Y, emulsifier: soya lecithin, Z.

These labelling rules will remain in force until 13 December 2014 and will therefore run in parallel with the Food Information Regulation (EU) No. 1169/2011.[30] The new Regulation requires allergenic ingredients to be listed in the ingredients list in a typeset that clearly distinguishes it from the rest of the list of ingredients, for example by means of the font, style or background colour, *e.g.*: Ingredients: sugar, **butter**, **eggs**, **wheat** flour, emulsifier: **soya** lecithin, flavouring. In the absence of an ingredients list the word "contains" followed by the allergenic substance is required on the label.

4.8.3   Specific Labelling for Six Food Colours

Following on from research, a labelling requirement relating to hyperactivity in children has been introduced for six colours under Annex V of Regulation (EC) No. 1333/2008 (as amended) on food additives.[8] These colours are:

- E110 (sunset yellow FCF);
- E104 (quinoline yellow);
- E122 (carmoisine);
- E129 (allura red);
- E124 (ponceau 4R); and
- E102 (tartrazine).

Table 4.4 List of EU allergens.

---

1. Cereals containing gluten, namely: wheat, rye, barley, oats, spelt, kamut or their hybridised strains, and products thereof, except:
   (a) wheat-based glucose syrups including dextrose (1);
   (b) wheat-based maltodextrins (1);
   (c) glucose syrups based on barley;
   (d) cereals used for making alcoholic distillates including ethyl alcohol of agricultural origin
2. Crustaceans and products thereof
3. Eggs and products thereof
4. Fish and products thereof, except:
   (a) fish gelatine used as carrier for vitamin or carotenoid preparations
   (b) fish gelatine or Isinglass used as fining agent in beer and wine
5. Peanuts and products thereof
6. Soybeans and products thereof, except:
   (a) fully refined soybean oil and fat (1)
   (b) natural mixed tocopherols (E306), natural D-alpha tocopherol, natural D-alpha tocopherol acetate, and natural D-alpha tocopherol succinate from soybean sources
   (c) vegetable oils derived phytosterols and phytosterol esters from soybean sources
   (d) plant stanol ester produced from vegetable oil sterols from soybean sources
7. Milk and products thereof (including lactose), except:
   (a) whey used for making alcoholic distillates including ethyl alcohol of agricultural origin
   (b) lactitol
8. Nuts, namely: almonds (*Amygdalus communis* L.), hazelnuts (*Corylus avellana*), walnuts (*Juglans regia*), cashews (*Anacardium occidentale*), pecan nuts (*Carya illinoinensis* (Wangenh.) K. Koch), Brazil nuts (*Bertholletia excelsa*), pistachio nuts (*Pistacia vera*), macadamia or Queensland nuts (*Macadamia ternifolia*), and products thereof, except for nuts used for making alcoholic distillates including ethyl alcohol of agricultural origin
9. Celery and products thereof
10. Mustard and products thereof
11. Sesame seeds and products thereof
12. Sulfur dioxide and sulfites at concentrations of more than 10 mg/kg or 10 mg/l in terms of the total SO_2 that are to be calculated for products as proposed ready for consumption or as reconstituted according to the instructions of the manufacturers
13. Lupin and products thereof
14. Molluscs and products thereof

---

(1) and the products thereof, in so far as the process that they have undergone is not likely to increase the level of allergenicity assessed by the authority for the relevant product from which they originated.

From 20 July 2010, the following declaration:

> *"name or E number of the colour(s): may have
> an adverse effect on activity and attention in children"*

has to be included on the label of food products containing any of the above food colours. Any food product placed on the market before 20 July 2010 that does not comply with this labelling requirement will be able to be marketed until its date of minimum durability or use-by date expires.

The above labelling requirement does not apply:

- where the colours have been used for health or other marking of meat or the stamping or other decoration of eggshells;
- to beverages containing more than 1.2% by volume of alcohol.[33]

References

1. Regulation (EC) No. 178/2002 of the European Parliament and of the Council of 28 January 2002 laying down the general principles and requirements of food law, establishing the European Food Safety Authority and laying down procedures in matters of food safety L031, 01.02.2002, 1–24.
2. Council Directive 89/107/EEC of 21 December 1988 on the approximation of the laws of the Member States concerning food additives authorised for use in foodstuffs intended for human consumption L040, 11.02.1989, 27–33.
3. European Parliament and Council Directive 94/35/EC of 30 June 1994 on sweeteners for use in foodstuffs L237, 10.09.1994, 3–12.
4. European Parliament and Council Directive 94/36/EC of 30 June 1994 on colours for use in foodstuffs L237, 10.09.1994, 13–29.
5. European Parliament and Council Directive 95/2/EC of 20 February 1995 on food additives other than colours and sweeteners L61, 18.03.1995, 1–40.
6. Regulation (EC) No. 1331/2008 of the European Parliament and of the Council of 16 December 2008 establishing a common authorisation procedure for food additives, food enzymes, and food flavourings L354, 31.12.2008, 1–6.
7. Regulation (EC) No. 1332/2008 of the European Parliament and of the Council of 16 December 2008 on food enzymes and amending Council Directive 83/417/EEC, Council Regulation (EC) No. 1493/1999, Directive 2000/13/EC, Council Directive 2001/112/EC and Regulation (EC) No. 258/97 L354, 31.12.2008, 7–15.
8. Regulation (EC) No. 1333/2008 of the European Parliament and of the Council of 16 December 2008 on food additives L354, 31.12.2008, 16–33.
9. Regulation (EC) No. 1334/2008 of the European Parliament and of the Council of 16 December 2008 on flavourings and certain food ingredients with flavouring properties for use in and on foods and amending Council Regulation (EEC) No. 1601/91, Regulations (EC) No. 2232/96 and (EC) No. 110/2008 and Directive 2000/13/EC L354, 31.12.2008, 34–50.
10. Commission Regulation (EU) No. 1129/2011 of 11 November 2011 amending Annex II to Regulation (EC) No. 1333/2008 of the European Parliament and of the Council by establishing a Union list of food additives L295, 12.11.2011, 1–177.
11. Commission Regulation (EU) No. 1130/2011 of 11 November 2011 amending Annex III to Regulation (EC) No. 1333/2008 of the European Parliament and of the Council on food additives by establishing a Union

list of food additives approved for use in food additives, food enzymes, food flavourings and nutrients L295, 12.11.2011, 178–204.

12. Commission Directive 2008/60/EC of 17 June 2008 laying down specific purity criteria concerning sweeteners for use in foodstuffs (Codified version) L158, 18.06.2008, 17–40.

13. Commission Directive 2008/128/EC of 22 December 2008 laying down specific purity criteria concerning colours for use in foodstuffs (Codified version) L6, 10.01.2009, 20–63.

14. Commission Directive 2008/84/EC of 27 August 2008 laying down specific purity criteria on food additives other than colours and sweeteners L253, 20.09.2008, 1–175.

15. Commission Regulation (EU) No. 231/2012 of 9 March 2012 laying down specifications for food additives listed in Annexes II and III to Regulation (EC) No. 1333/2008 of the European Parliament and of the Council L83, 22.03.2012, 1–295.

16. Council Directive 98/83/EC of 3 November 1998 on the quality of water intended for human consumption L330, 05.12.1988, 32–54.

17. Directive 2009/32/EC of the European Parliament and of the Council of 23 April 2009 on the approximation of the laws of the Member States on extraction solvents used in the production of foodstuffs and food ingredients (Recast) L141, 06.06.2009, 3–11.

18. Commission's administrative guidance for the request of authorisation of a food additive, 23.07.2009, http://ec.europa.eu/food/food/chemicalsafety/additives/flav16_en.pdf.

19. Commission Regulation (EU) No. 234/2011 of 10 March 2011 implementing Regulation (EC) No. 1331/2008 of the European Parliament and of the Council establishing a common authorisation procedure for food additives, food enzymes and food flavourings L64, 11.03.2011, 15–24.

20. Food Additives, Enzymes and Flavourings - Applying for an authorisation, 29.11.2011, http://ec.europa.eu/food/food/fAEF/authorisation_application_en.htm.

21. Regulation (EC) No. 1829/2003 of the European Parliament and of the Council of 22 September 2003 on genetically modified food and feed L268, 18.10.2003, 1–23.

22. Regulation (EC) No. 258/97 of the European Parliament and of the Council of 27 January 1997 concerning novel foods and novel food ingredients L43, 14.02.1997, 1–6.

23. Commission Regulation (EU) No. 1131/2011 of 11 November 2011 amending Annex II to Regulation (EC) No. 1333/2008 of the European Parliament and of the Council with regard to steviol glycosides L295, 12.11.2011, 205–211.

24. Commission Regulation (EU) No. 232/2012 of 16 March 2012 amending Annex II to Regulation (EC) No. 1333/2008 of the European Parliament and of the Council as regards the conditions of use and the use levels for Quinoline Yellow (E 104), Sunset Yellow FCF/Orange Yellow S (E 110) and Ponceau 4R, Cochineal Red A (E 124) L78, 17.03.2012, 1–12.

25. Commission Regulation (EU) No. 380/2012 of 3 May 2012 amending Annex II to Regulation (EC) No. 1333/2008 of the European Parliament and of the Council as regards the conditions of use and the use levels for aluminium-containing food additives L119, 04.05.2012, 14–38.

26. Commission Regulation (EU) No. 470/2012 of 4 June 2012 amending Annex II to Regulation (EC) No. 1333/2008 of the European Parliament and of the Council as regards the use of polydextrose (E 1200) in beer L144, 05.06.2012, 16–18.

27. Commission Regulation (EU) No. 471/2012 of 4 June 2012 amending Annex II to Regulation (EC) No. 1333/2008 of the European Parliament and of the Council as regards the use of lysozyme (E 1105) in beer L144, 05.06.2012, 19–21.

28. Commission Regulation (EU) No. 472/2012 of 4 June 2012 amending Annex II to Regulation (EC) No. 1333/2008 of the European Parliament and of the Council as regards the use of glycerol esters of wood rosins (E 445) for printing on hard-coated confectionery products L144, 05.06.2012, 22–24.

29. Directive 2000/13/EC of the European Parliament and of the Council of 20 March 2000 on the approximation of the laws of the Member States relating to the labelling, presentation and advertising of foodstuffs L109, 06.05.2000, 29–42.

30. Regulation (EU) No. 1169/2011 of the European Parliament and of the Council of 25 October 2011 on the provision of food information to consumers, amending Regulations (EC) No. 1924/2006 and (EC) No. 1925/2006 of the European Parliament and of the Council, and repealing Commission Directive 87/250/EEC, Council Directive 90/496/EEC, Commission Directive 1999/10/EC, Directive 2000/13/EC of the European Parliament and of the Council, Commission Directives 2002/67/EC and 2008/5/EC and Commission Regulation (EC) No. 608/2004 L304, 22.11.2011, 18–63.

31. European Commission's Food Additive Database https://webgate.ec.europa.eu/sanco_foods/main/?event=display.

32. Commission Regulation (EC) No. 884/2007 of 26 July 2007 on emergency measures suspending the use of E 128 Red 2G as food colour L195, 27.07.2007, 8–9.

33. Commission Regulation (EU) No. 238/2010 of 22 March 2010 amending Annex V to Regulation (EC) No. 1333/2008 of the European Parliament and of the Council with regard to the labelling requirement for beverages with more than 1,2% by volume of alcohol and containing certain food colours L75, 23.03.2010, 17–17.

CHAPTER 5

Legislation for Food Additives Outside Europe

VANESSA RICHARDSON,*[a] ELLA FREEMAN,[c]
LAURA FITZPATRICK,[a] LINDA AMIRAT,[b]
MARIKO KUBO[a] AND MENG LI[c]

[a] Principal Regulatory Advisor, Leatherhead Food Research, UK;
[b] Regulatory Advisor, Leatherhead Food Research, UK; [c] Senior Regulatory
Advisor, Leatherhead Food Research, UK
*E-mail: Legislation@LeatherheadFood.com

5.1 Introduction

Vanessa Richardson, Principal Regulatory Advisor, Leatherhead Food Research

A major problem for manufacturers of food products, importers and exporters over the years has been that of trying to find ingredient and additive specifications and conditions of use that will enable a particular food product of interest to be sold in more than one country. The world is becoming smaller as many companies now look to sell their products in overseas markets and many choose to purchase ingredients from outside the European Union (EU). However, additives legislation can differ significantly from country to country, both in the way it is structured and in details such as the acceptability of named additives in individual foods. It is true that developments in key trading blocs such as the EU or international standards, such as those developed by the Codex Alimentarius Commission (CAC) have influenced additives legislation in other parts of the world. The differences in the manner

Essential Guide to Food Additives, 4th Edition
Edited by Mike Saltmarsh
© The Royal Society of Chemistry 2013
Published by the Royal Society of Chemistry, www.rsc.org

in which food additives are regulated are still a major concern to both ingredients and additives suppliers and also to manufacturers of the final food product.

Two factors that strongly influence the use of particular food additives are technological need and safety. It is necessary to ensure that levels of particular additives do not increase above acceptable safety limits for those people who, by nature of their diet, may consume high amounts of a particular additive, for example by the consumption of significant amounts of a sweetener through consumption of large quantities of soft drinks. The establishment of Acceptable Daily Intakes (ADIs) for additives is referred to throughout the chapter.

5.1.1   Labelling Issues

Another factor to consider is labelling. There was a trend some years ago to replace chemical names on a label with equivalent numbers – for example, E numbers. The pendulum has now swung the other way with the widespread popularity of clean labels, where manufacturers often prefer to use additive names wherever possible (taking account of the complexity of the chemical name and the limited room they may have on the label) and avoiding the use of numbers for reasons related to consumer perception linked to the use of food additives. Generally, today, a mixture of both is used. Consumers have a right to make an informed choice about the foods they consume, and most additives are required to be declared on the label as part of the ingredients list. However, those that are present only as a result of carry-over from inclusion in another ingredient, with no technological effect in the final food, or those used as processing aids, are normally exempted from declaration unless they, or the raw materials from which they are derived, are known to cause allergies or intolerances. Exact rules for exemptions apply.

Compounds added for their nutritional benefit are not normally regarded as food additives. At the same time an additive used in a food for a technological purpose should not be declared as a nutrient, but with the category name of the function to which it relates. For example, if ascorbic acid is being added as an antioxidant, it should be declared as an antioxidant and not as vitamin C. However, different rules may apply in different countries.

Another factor of relevance is that of purity criteria, *i.e.* the specification to which additives must be manufactured. Different specifications may apply in the USA (established by the Food Chemicals Codex (FCC)), in the EU, in countries that follow the Codex Alimentarius and in certain individual countries that establish their own criteria.

In this chapter, developments in food additives at Codex level are assessed and key aspects of international food legislation in some of the major export nations outside Europe, *i.e.* the USA, Canada, Japan and certain Far East countries, the Southern Common Market (MERCOSUR), the Middle East and Australia/New Zealand, are discussed.

5.2 Codex Alimentarius

Vanessa Richardson, Principal Regulatory Advisor, Leatherhead Food Research

The Food and Agriculture Organization (FAO) of the United Nations, and the World Health Organization (WHO) established the Codex Alimentarius Commission (CAC) in 1963 after recognising a need for international industry standards for the food industry worldwide, in order to protect the health of consumers and to make international trade in food easier. To this end the CAC developed a "collection of standards, codes of practice, guidelines and other recommendations"[1] that are known as the Codex Alimentarius. However, the term *Codex* is often used to describe the Codex body (*i.e.* the CAC), and is occasionally used in this context in this Guide.

The CAC is composed of a secretariat, an executive committee and several subject-specific subsidiary bodies and task forces (including *ad hoc* committees). Any Member Nation and Associate Member of the FAO and WHO that is interested in international food standards may become a member of the CAC upon application. Non-members of the CAC with special interest in its work may attend CAC sessions, or sessions of its subsidiaries and *ad hoc* meetings as observers, provided they request to do so. The EU is a Member of the CAC, even though the individual Member States are also Members. Nations that are not members, but belong to the United Nations, may be invited, if they request, as observers.[2]

The aim of the Codex Alimentarius is to protect public health, ensure fair trading and promote harmonisation. The documents (standards, guidelines and codes of practice) are developed by consensus and on the best available scientific and technical advice. A uniform procedure has been established for the development of Codex documents, known as the "step" procedure. There are eight steps in the procedure. At step 8, a draft is submitted to the CAC, along with any proposals for amendment, with a view to adoption as a Codex document. An accelerated procedure can be followed, which is completed in five steps. Once published, Codex documents are sent to governments for acceptance and to international organisations with competence in the subject in question as named by the member governments.

There are Codex Alimentarius documents established in a variety of topics, including:

- Food composition;
- Labelling;
- Food additives;
- Hygiene and safety;
- Pesticides;
- Contaminants;
- Veterinary drugs.

In many cases, recommended uses for additives are contained within commodity-based compositional standards. Much of the work in producing commodity-based compositional standards is handled by commodity-specific subsidiary bodies (*e.g.* the Codex Committee on Fats and Oils (CCFO)), with final endorsement by the Codex Committee on Food Additives (CCFA), the subsidiary body primarily responsible for the work on food additives, and the CAC. The trend these days, however, continues to be away from so-called "product-specific" legislation to horizontal provisions aimed at all food types. This, and the increasingly important role of Codex documents as reference texts in trade disputes under the World Trade Organization (WTO) agreements, made it appropriate to establish a General Standard on Food Additives (GSFA), which was intended to serve as "the single authoritative reference point for food additives".[3] This follows the EU pattern of having additives and their food uses detailed together and separately from individual product standards.

The aim of establishing permitted levels of use of additives in various food groups is to ensure that the intake of additives does not exceed the Acceptable Daily Intake (ADI).

5.2.1   Codex General Standard on Food Additives

The Codex General Standard for Food Additives (GSFA) was adopted as a Codex standard in 1995 (CODEX STAN 192-1995). Since then, there have been several amendments to this Standard, with the latest being adopted in 2011. Only the additives that have been evaluated as safe by the Joint FAO/WHO Expert Committee on Food Additives (JECFA) and have been assigned an ADI and an International Numbering System (INS) designation by Codex are considered for inclusion in this Standard. The Standard gives details of conditions under which permitted food additives may be used.

The GSFA was intended to serve as "the single authoritative reference point for food additives" and additive provisions within it were supposed to match those given in commodity-based compositional standards. However, some of the information contained within the GSFA contradicts that in commodity-based compositional standards and it is still not clear which standard should take precedence as different national authorities worldwide have diverging views on the subject.

According to the GSFA, a food additive is defined as:[4]

Any substance not normally consumed as a food by itself and not normally used as a typical ingredient of the food, whether or not it has nutritive value, the intentional addition of which to food for a technological (including organo-leptic) purpose in the manufacture, processing, preparation, treatment, packing, packaging, transport or holding of such food results, or may be reasonably expected to result (directly or indirectly), in it or its by-products becoming a component of or otherwise affecting the characteristics of such foods. The term does not include contaminants or substances added to food for maintaining or improving nutritional qualities.

According to the preamble of the GSFA,[4] the use of food additives is justified only when such use:

- Has an advantage for;
- Does not present a hazard to the health of;
- Does not mislead the consumer; and
- Serves one or more of the purposes and needs established below and only when these cannot be achieved by other means that are economically and technologically practicable:
  a. To preserve the nutritional quality of the food; an intentional reduction in nutritional quality of a food would be justified in (b) below and in other circumstances where the food does not constitute a significant item in a normal diet;
  b. To provide necessary ingredients or constituents for foods manufactured for groups of consumers having special dietary needs;
  c. To enhance the keeping quality or stability of a food or to improve its organoleptic properties, provided this does not change the nature, substance or quality of the food so as to deceive the consumer;
  d. To provide aids in the manufacture, processing, preparation, treatment, packaging, transport or storage of food, provided the additive is not used to disguise the effects of use of faulty raw materials or undesirable (including unhygienic) practices or techniques during the course of any of these activities.

The food category system is hierarchical, meaning that when an additive is permitted in a general category, it is also permitted in all its subcategories, unless otherwise stated. It is based on food category descriptors as marketed and is also in compliance with the carry-over principle. An additive may be acceptable in a final food provided it is permitted in one of the component ingredients or raw materials according to this Standard and its amount in these does not exceed the maximum permitted, and provided that the amount of that ingredient present in the final food will not be higher than it would be by the use of the ingredients under proper technological conditions or manufacturing practices. It is used to simplify the reporting of food additive uses for the development of this Standard.[4]

The food additives considered were grouped into 27 major functional classes, in accordance with the Codex Guidelines on Class Names and the International Numbering System for Food Additives.[5]

The GSFA is developed on an additive-by-additive basis rather than by functional class. This Standard contains a detailed list of additives, or food additive groups, *e.g.* phosphates, permitted for use under specified conditions in certain food categories, or individual food items – for example, Caramels III and IV as colours in named foods, ferric ammonium citrate as an anticaking agent in named foods, and polydimethylsiloxane as an anticaking, antifoaming agent or emulsifier in named foods (given in Table 1 of the GSFA). Table 2 contains the same information as Table 1, but it is arranged by food category number. Table 3 of the Standard has a list of additives permitted for food use in general, unless otherwise specified, in accordance with the principles of Good Manufacturing Practice (GMP).

According to GMP:[4]

- *The quantity of additive added to food shall be limited to the lowest possible level necessary to accomplish its desired effect;*
- *The quantity of the additive that becomes a component of food as a result of its use in the manufacturing, processing or packaging of a food and which is not intended to accomplish any physical, or other technical effect in the food itself, is reduced to the extent reasonably possible;*
- *The additive is of appropriate food grade quality and is prepared and handled in the same way as a food ingredient.*

This list includes a number of additives across a range of technical functions – for example, guar gum, beet red, papain and polydextrose. However, although generally permitted in accordance with GMP, it is recognised that it is not in line with GMP to use additives in a number of unprocessed foods or basic foods without significantly changing the nature of the product. The use of Table 3 additives is restricted in products such as plain milk and buttermilk; fresh fruit and vegetables; liquid and frozen egg products; honey; fats and oils that are essentially free of water; coffee, tea and herbal infusions; infant and follow-on formula; and fresh or dried pastas and noodles.

5.2.2 JECFA

Risk analysis is key to the work of the CAC. It is the role of the Joint FAO/WHO Expert Committee on Food Additives (JECFA), the risk assessor of the Joint FAO/WHO Food Standards Programme, to evaluate food additives in terms of their toxicology and to prepare specifications for each additive, including, where necessary, an ADI, examples of ADIs are provided in Table 5.1. Independent expert scientific advice is provided at an international level. Additives are only one interest of JECFA; it also deals with contaminants, veterinary residues and naturally occurring toxicants.

Table 5.1 ADIs of some commonly used sweeteners – JECFA.

Additive	ADI (mg/kg body weight)
Calcium cyclamate	0–11(expressed as cyclamicacid). Group ADI for calcium and sodium salts
Aspartame	0–40 for aspartame and 0–7.5 for diketopiperazine
Acesulfame K	0–15
Saccharin	0–5 Group ADI for saccharin and Ca, K and Na salts

An ADI in this context is an estimate of the amount of a food additive in food or drinking water, expressed on a bodyweight basis, that can be taken daily in the diet, over a lifetime, without appreciable health risk. The weight of a standard man is taken as 60 kg. Generally, ADIs are expressed in mg/kg bodyweight, in a range from 0 to an upper limit, which is considered to be the zone of acceptability of the substance. The acceptable level established is an upper limit and the Committee encourages the lowest levels of use that are technologically feasible.

The ADI may be qualified at present by a number of terms. In the context of food additives, the following definitions apply:[6]

Not specified – If an ADI is not specified, then, on the basis of available chemical, biochemical, toxicological and other data, the total dietary intake of the substance arising from use at levels necessary to achieve the desired effect and from its acceptable background in food, does not, in JECFA's opinion, represent a hazard to health. A numerical ADI is not, therefore, deemed necessary. An additive with a nonspecified ADI must be used in accordance with GMP principles.

Not limited – A term that is no longer used by JECFA with the same meaning as *not specified*.

Conditional – A term no longer used by JECFA, but that was used to signify a range above the *unconditional ADI*, which may signify the acceptable intake when special circumstances, different patterns of dietary intake or special groups of population require consideration.

Temporary – A temporary ADI may be allocated when sufficient data are available to conclude that the use of the substance is safe over the relatively short period of time needed to generate and evaluate additional safety data, but are insufficient to ensure that the substance is safe over a lifetime of usage. A higher-than-normal safety factor is used when a temporary. A higher-than-normal safety factor is used when a temporary ADI is established and a deadline is set so that appropriate data can be submitted to JECFA to resolve the safety issue.

Not allocated – There are several instances when an ADI is not allocated, including lack of information on a particular substance, data on adverse effects calling for advice that the substance should not be used, *etc*.

Group ADI – Additives that are closely related chemically and toxicologically can be grouped together for the purposes of evaluation, for example polyoxyethylene sorbitan esters and modified celluloses. The ADI is expected to cover all the members of the group that may be included in the diet.[7]

5.2.3 Specifications

Specifications for the identity and purity of food additives developed by the Committee have three purposes:[8]

- To ensure that the additive has been biologically tested;

- To ensure that the substance is of the quality required for the safe use in food;
- To reflect and encourage Good Manufacturing Practice.

Specifications include additive synonyms, definition, assay, description, functional uses, tests of identity and impurities and assay of major components. The specifications are periodically reviewed, due to changes in patterns of additive use, in raw materials and in methods of manufacture, as well as in light of new scientific data. Specifications may be full or tentative; *tentative* is used only in cases where the available data is insufficient to establish a full specification. In such cases JECFA states the level of information that is necessary and sets a date by which this information must be provided. Once all missing data is received, and provided that this data is sufficient, the *tentative* designation is removed and the specification given full status.

Additives in accordance with the GSFA should be of appropriate food-grade quality and conform with the specification recommended by Codex, *i.e.* those set by JECFA and duly approved by Codex (as per the List of Codex Specifications for Food Additives, CAC/MISC 6-011),[9] or, in the absence of such specifications, with appropriate specifications developed by responsible national or international bodies. Food-grade quality is achieved by compliance with the specification as a whole and not merely with individual criteria in terms of safety.

JECFA meetings are usually held annually, as are CCFA meetings; the list of substances scheduled to be evaluated and requests for data are normally issued in advance of these meetings.

5.3   Food Additives Legislation in Other Countries

5.3.1   USA

Laura Fitzpatrick, Principal Regulatory Advisor, Leatherhead Food Research

Food additives legislation in the USA has evolved in a unique manner and it is appropriate to consider how it differs from that applying elsewhere.

5.3.1.1   Framework

The Federal Food, Drug and Cosmetic Act (FFDCA)[10] lays down the framework for food safety at a Federal level in the USA. This includes the definitions and principles on the use of food additives. The provisions of the Act are enforced by the Food and Drug Administration (FDA) through more detailed regulations laid down in Title 21 of the Code of Federal Regulations[11] (21 CFR). The FDA is responsible for all food products including dietary supplements, and all food additives; however, other agencies have specific responsibilities for eggs, meat, poultry and alcoholic beverages.

Despite this, the overall framework for the regulation of food additives is the same. The US Department of Agriculture (USDA) is responsible for meat and poultry products, and provisions on permitted additives in these products are laid down in Title 9 of the Code of Federal Regulations (9 CFR). The Alcohol and Tobacco Tax and Trade Bureau (TTB) is responsible for alcoholic beverages, and permitted additives for these products are laid down in Title 27 of the Code of Federal Regulations (27 CFR).

The Food Additives Amendment[10] was enacted to the FFDCA in 1958 and forms the present basis of the regulation of food additives. This Amendment defined the terms *food additive* and *unsafe food additive* and established a premarket approval process for food additives. The adulteration provisions were amended to deem any food that contains any substance not regulated by these provisions as unsafe.

In accordance with the Act, the term *food additive* means "any substance of which the intended use results, or may reasonably be expected to result, directly or indirectly, in its becoming a component of any food, or otherwise affecting the characteristics of the food." This includes any substance intended for use in manufacturing, processing, treating or holding food, and any source of radiation used.

In enacting the Amendment, the Congress recognized that many substances that were intentionally added to food would not require a formal premarket review by the FDA to assure their safety. This was either because their safety had already been established by a long history of use in food or by virtue of the nature of the substances and the information generally available to scientists about the substances. As a result, the definition of a food additive excludes substances that are Generally Recognized as Safe (GRAS), under the conditions of their intended use, among experts qualified by scientific training and experience to evaluate their safety. The view that a substance is GRAS may be based on scientific procedures or, for substances used in food prior to January 1958, on experience derived from its common use in food. Colour additives are also excluded from the definition of a food additive and there are separate provisions on these laid down in the Act.

Therefore, substances intended for use in the manufacture of foodstuffs for human consumption are classified into three categories: food additives, prior-sanctioned food ingredients and substances Generally Recognized as Safe (GRAS); examples are provided in Table 5.2. In addition, there are separate provisions on colour additives. These will be described in turn.

21 CFR Parts 170–189 lay down regulations on these food additives and GRAS substances in detail, including the procedures for their approval,

Table 5.2 Additives classification in the USA.

Class	Example
Food additive	Saccharin
Prior-sanctioned food additive	Sodium nitrate
GRAS substance	Sorbitol

labelling requirements, specifications and purity criteria. In order to clarify the provisions of their use, 43 general food categories and 32 physical or technical functions have been established.

5.3.1.2 Food Additives

A food additive may be direct, secondary direct or indirect, depending on how it is used.

Direct food additives are divided into eight categories: food preservatives; coatings, films and related substances; special dietary and nutritional additives; anticaking agents; flavouring agents and related substances; gums, chewing-gum bases and related substances; other specific usage additives; and multi-purpose additives.

Secondary direct food additives are components used in ingredients of processed foods that may become additives in the final food. These are divided into four categories: polymer substances and polymer adjuvants for food treatment; enzyme preparations and micro-organisms; solvents, lubricants, release agents and related substances; and specific usage additives.

Some direct food additives are listed separately as being permitted on an interim basis pending additional study. This listing applies when new information raises substantial questions about the safety or functionality of a substance, although there is reasonable assurance that no harm to public health will result from continued use of the substance while further study is carried out. Substances listed in this section include saccharin and its salts, and brominated vegetable oil.

Indirect food additives are materials that may become part of a food as a component of packaging material, adhesives, food-processing equipment, surfaces and containers used for food handling, and certain production aids and sanitisers. Substances used in food-contact articles that may be expected to migrate into food at negligible levels may be exempted from food additive status by petition. However, this would not be necessary if the substance was considered GRAS, or GRAS for use in food packaging, or was a substance that had prior-sanction approval (see below).

Food additives are approved for use by a petition procedure. Details of this are laid down in the legislation. The FFDCA requires that a regulation regarding the use of a food additive is issued 90 days after the petition is filed unless the time is extended to 180 days. However, in practice, food-additive petitions can take much longer than this as the clock stops every time new information is required. The FDA has an expedited review procedure for certain food-additive petitions that are expected to have a significant impact on food safety. These petitions are placed at the beginning of appropriate review queues and apply to additives that are intended to decrease the incidence of foodborne illness through antimicrobial action against human pathogens or their toxins in or on food.[12]

Guidance for industry regarding the petitioning procedure and details of pending food additive and colour additive petitions are available via the FDA Internet site.[13]

5.3.1.3 Prior-Sanctioned Ingredients

Prior-sanctioned ingredients listed in the legislation are substances that received official approval for their use in food by the FDA or the USDA prior to the Food Additives Amendment in 1958.[10] All the substances listed in this section are components of food-packaging materials, with the exception of sodium nitrate and potassium nitrate.

5.3.1.4 Substances that are Generally Recognised as Safe (GRAS)

21 CFR 182 lists some substances that were used in food prior to 1958 without known detrimental effects, whose regulatory status was clarified by the FDA. The regulations specifically state that it is impractical to list all substances that are GRAS for their intended use, such as common food ingredients and monosodium glutamate.

21 CFR 184 details substances that have been affirmed as GRAS for particular purposes after review by the FDA. However, this list does not contain all GRAS substances as the responsibility for proof of safety lies with the additive manufacturer. The list of substances that have been affirmed as GRAS includes many substances that are widely seen as food additives, such as locust bean gum, ammonium sulfate and adipic acid; in contrast, many substances that are typical food ingredients are also listed, such as garlic, sucrose and malt syrup.

It is the use of a substance, rather than the substance itself, that is eligible for GRAS exemption. This means that an affirmation may be granted for a particular use without taking into account other uses that may also be GRAS.

A determination that a particular use of a substance is GRAS requires both technical evidence of its safety and a basis to conclude that this evidence is generally known and accepted. In contrast, a determination that a food additive is safe requires only technical evidence of safety. Therefore, a GRAS substance is distinguished from a food additive on the basis of the common knowledge about the safety of the substance for its intended use rather than on the basis of what the substance is or the types of data and information that are necessary to establish its safety. In addition, the FDA has pointed out that the existence of a severe conflict among experts regarding the safety of the use of a substance precludes a finding of general recognition.[14]

It is important to note that the GRAS exemption applies to the premarket approval requirements for food additives only and there is no corresponding exemption to the premarket approval requirements for colour additives. In other words, no colour additive in the USA has GRAS status.

Until 1998, there was a procedure in effect by which manufacturers could petition the FDA to affirm that a substance is GRAS under certain conditions of use. This petition process provided a mechanism for official recognition of GRAS determinations. GRAS petitions were in the public domain, whereas additives petitions are not, and unlike additive petitions, there was no time limit laid down in the legislation for the petition.

The FDA recognized that the petition process took up significant amounts of its resources and, on 17 April 1997, it issued a proposal to replace this procedure and to clarify when the use of a substance is GRAS.[14] The proposed rule was intended to replace the GRAS affirmation process with a notification procedure. This means that any person could notify the FDA of a determination that the use of a particular substance is GRAS. Within 90 days of receipt of the notice, the FDA would respond to the notifier in writing and advise the notifier that the Agency had identified a problem with the notice or otherwise. However, the FDA would not conduct its own detailed evaluation of the data that the notifier relies on to support a determination that a use of a substance is GRAS nor would it affirm that the substance was GRAS for its intended use. In the interim between this proposal and any final rule, the FDA has encouraged use of this procedure. At the time of printing, a Final Rule has not been issued, but the GRAS notification program is still active "in the interim".

Since the issuing of the proposed rule, the FDA has received and responded to over 400 of such GRAS notices for a variety of substances, and has published an inventory of these notices and the Agency's response. This is available on the Internet and is up dated approximately on a monthly basis.[15]

5.3.1.5   *Examples of Current GRAS Notifications on the Internet*

Among the notifications currently available for viewing on the website are those for a number of enzymes (including 1,4-α-glucan branching enzymes from named sources, and a peroxidase enzyme preparation from a genetically modified *Aspergillus niger*), two strains of *Lactobacillus reuteri*, one for use in multiple foods and beverages and one in powdered whey-based infant formula, sulfate-buffered sulfuric acid as an antimicrobial on meat and poultry, erythritol, several purified steviol glycosides and Rebaudioside A notifications and various phytosterol and stanol formulations. These illustrate the range of compounds for which GRAS notification can be given. Some of the compounds are more *traditional* additives and others would be more likely to be classified as ingredients in Europe, albeit with specific approval required as novel ingredients in some cases.

5.3.1.6   *Example of Additive Approval – Olestra*

Another example of differences in classification is that of Olestra. Under European law, approval for Olestra was initially sought as a novel food in the UK prior to the current authorisation scheme under the EU novel foods Regulation, Regulation (EC) No. 258/97,[16] rather than falling into the scope of additives legislation. This is due to additive regulation in the European Union being defined by the function of the compound in question. The mode of action of Olestra takes the compound outside the scope of the additives legislation.

Title 21 172.867 of the Code of Federal Regulations defines Olestra as a mixture of octa-, hepta- and hexa-esters of sucrose with fatty acids derived from edible fats and oils or fatty acid sources that are generally recognised as safe or approved for use as food ingredients. Various specifications are laid down. Olestra is authorised for use in place of fats and oils in prepackaged ready-to-eat savoury snacks (*i.e.* salty or piquant but not sweet) and prepackaged, unpopped popcorn kernels that are ready-to-heat. In such foods, the additive may be used in place of fats and oils for frying or baking, in dough conditioners, in sprays, in filling ingredients or in flavours. To compensate for any interference with absorption of fat-soluble vitamins, alpha-tocopherol, retinole quivalent, vitamin D and vitamin K have to be added in quantities as specified. The label of a food containing Olestra must carry the following information:

The added vitamins must be included in the list of ingredients, but are not considered in determining nutrient content for nutrition labelling or any nutrition claims, express or implied. An asterisk shall follow vitamins A, D, E, and K in the listing of ingredients, appearing as a superscript following each vitamin. Immediately following the ingredient list an asterisk and statement, "Dietarily insignificant" shall appear prominently and conspicuously.

In order to be consistent with its obligation to monitor the safety of all food additives, the FDA stated it would review and evaluate all data and information bearing on the safety of Olestra received by the Agency after the Regulation came into force. Such data, information and evaluation would then be presented to the Agency's Food Advisory Committee within 30 months of the effective date of the Regulation. This occurred at an open public meeting, held June 15–17, 1998, in which new data and information concerning Olestra, obtained since the 1996 approval were presented. One outcome of this meeting, in conjunction with a petition from a food manufacturer, was that the FDA amended 172.867 to remove a requirement for a warning statement concerning the possible undesirable effects of Olestra on the gastrointestinal system, and its effect on the absorption of some vitamins and other nutrients.

5.3.1.7 Flavours

Flavours may be either food additives or GRAS. Some of these are specifically laid down in the food additives and GRAS provisions. In addition, the US Flavor and Extract Manufacturers' Association (FEMA) also publishes its own list of flavouring agents that it considers GRAS for food use – the "FEMA GRAS™ list".[17]

While the legislation does not specifically state that all flavouring agents appearing on the FEMA GRAS list are approved for food use, the FEMA expert panel is widely recognised as complying with the GRAS exemption, which requires the safety of a substance to be evaluated by experts that are experienced and trained in evaluating the safety of food substances.

5.3.1.8 Colours

Legislation for food additives used as colours in the USA is especially strict. All substances that are deliberately used for their colouring effect are colour additives. The FFDCA allows the use of approved colours only for this purpose. Provisions on specifically permitted colours are laid down in 21CFR Parts 73 and 74.

All colour additives are classed as *artificial*. They are divided into those that are exempt from certification and those that are subject to certification. All synthetic organic colours are subject to certification. This means that a sample from each batch is sent to the FDA, which determines the purity of the batch by laboratory analysis. Other colour additives are listed as those that are exempt from certification.

Colours that are subject to certification include FD&C blue no. 1 (brilliant blue FCF), FD&C blue no. 2 (indigo carmine), FD&C green no. 3 (fast green FCF), FD&C red no. 3 (erythrosine), FD&C red no. 40 (allura red), FD&C yellow no. 5 (tartrazine) and FD&C yellow no. 6 (sunset yellow FCF). Other colours are listed provisionally and await re-evaluation. This includes the lakes of these colours (except for the lake of FD&C red no.3).

Colours that are not subject to certification include annatto extract, dehydrated beets, canthaxanthin, caramel, β-apo-8'-carotenal, β-carotene, cochineal extract, carmine, toasted partially defatted cooked cottonseed flour, grape colour extract, fruit juice, vegetable juice, carrot oil, paprika, paprika oleoresin, riboflavin, saffron, titanium dioxide, turmeric and turmeric oleoresin. Other substances are listed but these are restricted to specific uses only, such as animal feeds.

As for food additives, approval for colour additives is by petition to the FDA. A specific procedure for colour additive petitions is laid down in the legislation.

It is important to note that foodstuffs for which a standard is laid down may only contain colour additives if specifically permitted by that standard. Nonstandardised foodstuffs, in general, may be coloured with permitted colour additives. However, the use of colour additives is not permitted if it conceals damage or inferiority, or if it makes the product appear better or of greater value than it is.

5.3.1.9 *Labelling*

Specific provisions on the labelling of some food additives and GRAS substances are laid down. Labelling provisions lay down general provisions on the labelling of colour additives and flavours.

In contrast to the rules in many other countries, most additive functions do not need to be declared in a list of ingredients. However, the function of all chemical preservatives, leavening agents and firming agents must be stated. Warning statements are also required if specific additives are used. A warning statement for products containing saccharin has been laid down in the Act. The FDA has established other warnings for some additives, such as aspartame and sorbitol. For sorbitol, the label must declare "excess consumption may have a laxative effect"[11] if an amount of more than 50 g sorbitol is consumed on a daily basis.

5.3.2 Canada

Laura Fitzpatrick, Principal Regulatory Advisor, Leatherhead Food Research

It could be thought that Canada might base its additives legislation on that of its close neighbour, the USA, particularly with the introduction of the North American Free Trade Agreement (NAFTA). However, this is not the case.

The Food and Drugs Act[18] lays down the framework for food safety and labelling in Canada. These provisions are implemented through the Food and Drug Regulations,[19] which include compositional standards, labelling requirements and provisions on food additives. The legislative and regulatory responsibility resides with Health Canada.

A food additive is defined by the Food and Drug Regulations as "any substance the use of which results, or may reasonably be expected to result, in it or its by-products becoming a part of or affecting the characteristics of a food". It does not include nutritive materials that are used, recognised or sold as food ingredients; vitamins, minerals and amino acids other than those listed as additives by the regulations; spices, seasonings, flavouring preparations, essential oils, oleoresins and natural extractives; agricultural chemicals other than those listed as additives by the regulations; food packaging materials; and drugs authorised for administration to animals reared as food.

There is a list of permitted food additives and only substances on this list may be used. Additives are classed in accordance with their function as specified by the regulations.

Maximum levels laid down in the Food and Drug Regulations relate to the finished food. It is important to know which foods are covered by compositional standards, as conditions for the use of additives may apply to specific standardised food categories. For all other foodstuffs that are not covered by compositional standards, many additives will have a general limit that applies. These foodstuffs are described as "Unstandardized Foods", which are defined in Division 1 of the Food and Drug Regulations as "any food for which a standard is not prescribed".[19]

Where the limit prescribed for a food additive is stated to be GMP (Good Manufacturing Practice), the amount of the food additive added to a food in manufacturing and processing shall not exceed the amount required to accomplish the purpose for which that additive is permitted to be added to that food. Permission from the authorities is required to use an additive that is not listed.

Interim Marketing Authorisations (IMAs) are used to bridge the time between the completion of the scientific evaluation of additives and amending the regulations. Additives used in accordance with an IMA are exempt from the requirement to be listed in the regulations. Health Canada maintains a list of IMAs that are in force.[20]

Food additives are listed in Division 16 of the Food and Drug Regulations according to function. These functional categories include food additives that may be used as:

- Anticaking Agents;
- Bleaching, Maturing and Dough Conditioning Agents;
- Colouring Agents;
- Emulsifying, Gelling, Stabilising and Thickening Agents;
- Food Enzymes;
- Firming Agents;
- Glazing and Polishing Agents;
- Miscellaneous Food Additives;
- Sweeteners;
- pH Adjusting Agents, Acid-Reacting Materials and Water Correcting Agents;
- Preservatives;
- Sequestering Agents;
- Starch Modifying Agents;
- Yeast Foods Carrier or Extraction Solvents.

Although there are provisions allowing for some additives, such as colours to be declared by an optional common name (e.g. 'colour'), there is not a general requirement for additives to be declared by these category names in the ingredients list.

5.3.3   Japan

Mariko Kubo, Principal Regulatory Advisor, Leatherhead Food Research

Food additives legislation differs again in Japan. Instead of having just one list of permitted additives, as is often the case, there are two in Japan. The List of Designated Food Additives contains additives mainly produced by a chemical reaction (other than a degradation process).[21] Designated food additives, therefore, include nature-identical additives.

Some additives have specific restrictive conditions of use.[22] Others are generally permitted for food use, subject to any requirements specified in compositional standards. A list of synthetic flavourings is included. The second list is the List of Existing Food Additives,[23] which is a list of those so-called natural food additives currently used in Japan. Most of the existing additives on this list come with their own definition, which is often very precise. Care has to be taken when considering a synthetic equivalent of a natural compound, for example, β-carotene; the natural form may be regulated by the List of Existing Food Additives[23] but the synthetic form by the List of Designated Food Additives.[21]

The range of compounds included in the List of Existing Food Additives varies from those commonly recognised as food additives in other countries to compounds not generally regulated elsewhere. Examples of the former category include gum arabic (defined as a substance composed mainly of polysaccharides obtained from the secretion of acacia trees), chlorophyll, smoke flavourings, pectin and beet red. Examples of compounds included in the latter category include grapefruit seed extract (defined as a substance composed mainly of fatty acids and flavonoids obtained from grapefruit seeds) and purple yam colour (defined as a substance composed mainly of cyanidine acylglucosides obtained from yam tuberous roots). A number of enzymes are included in the list, including hemicellulases, pectinases, chitosanase and xylanase. Such additives are generally understood to be acceptable for food use unless otherwise restricted by the Standards for Use of Food Additives.[22]

In contrast, some additives detailed in the List of Designated Food Additives are permitted only in named foods to the maximum limits stated in the Standards for Use of Food Additives (for example, sorbic acid is permitted to max. 1.0 g/kg in syrups, and calcium disodium EDTA to max. 0.035 g/kg in canned or bottled nonalcoholic beverages). In other cases, where no restrictions are given, for example with citric acid and glycerol esters of fatty acids, it is understood that such additives are generally permitted unless restricted or prohibited by a standard of composition. As there are few compositional standards in a European sense, with standards mainly relating to hygiene, such additives tend to be generally acceptable. In some cases, named additives (for example, permitted colours) are not permitted in a range of basic foods, but are otherwise generally permitted. The List of Designated Food Additives includes compounds permitted as processing aids, flavouring (of the synthetic type) and as dietary supplements (fortification substances).

5.3.3.1   Labelling

Similarly to the requirements established in other countries, the majority of food additives need to be declared on the label of food products sold in Japan. Additives must be declared in descending order of weight immediately after the list of ingredients. Additives must be declared by the prescribed specific name unless otherwise specified. For some additives, such as additives used as colours, preservatives or antioxidants, the technological functions must be declared in addition to the specific name. In this case the specific name needs to be given in brackets. A limited number of generic terms (functional class names) may be used instead of the specific names, such as seasoning, flavouring, gum base and bittering agent if such declaration is specifically allowed.[24] Processing aids and fortification substances need not be declared unless otherwise stipulated. One interesting characteristic of additives labelling requirements in Japan is that the use of term "natural" or any equivalent term implying *naturalness* is not permitted for food additives; hence, the declaration *natural flavouring* is prohibited.

5.3.4   Other Far East Countries

Mariko Kubo, Principal Regulatory Advisor, Leatherhead Food Research
Meng Li, Senior Regulatory Advisor, Leatherhead Food Research

Food additives legislation in the Far East countries can vary significantly from country to country. Due to the economic importance of several of these markets, it is important to highlight some of these differences as they can have a severe impact on trade.

In Malaysia, detailed food standards established in the Food Regulations 1985,[25] as amended, control the use of additives. There are a number of lists of permitted additives, including colours, preservatives and food conditioners. This latter category covers emulsifiers, stabilisers, thickeners, solvents, acidity regulators, modified starches, gelling agents, enzymes, anticaking agents and antifoaming agents.

Additives in Singapore are also controlled by detailed food standards, laid down in the Food Regulations 1988, as amended (2011 version);[26] lists of permitted antioxidants, colours, emulsifiers and stabilisers, preservatives, anticaking agents, antifoaming agents, sweeteners, flavour enhancers, humectants, sequestrants, nutrient supplements and general-purpose food additives are included.

The use of additives in China is controlled by the National Food Safety Standard GB 2760-2011 Standard for the Use of Food Additives, as amended.[27] GB 2760 lists permitted additives and specifies the conditions under which the additives may be used. It also regulates the use of flavourings, processing aids and enzymes. Certain additives are generally permitted, which may be used in foodstuffs to a level of use in accordance with Good Manufacturing Practice (GMP), with the exception of certain foodstuffs specified in GB 2760. As a tool for assigning additive uses, there is a hierarchical food category system established in GB 2760. When an additive is recognised for use in a general category, it is recognised for use in all its subcategories, unless otherwise stated.

There is specific legislation laid down in Hong Kong for preservatives, antioxidants, colours and sweeteners, but currently, there are no specific provisions laid down for the use of other classes of additives in foods in Hong Kong. The competent authority makes reference to the Codex General Standard for Food Additives (GSFA, CODEX STAN 192-1995, as amended),[4] and those set by other countries, in determining whether a food additive is fit for human consumption. In Taiwan, additives are regulated by the Scope and Application Standards of Food Additives as amended,[28] which give positive lists of various additives including flavourings, solvents and substances for fortification. In Vietnam, the Ministry of Health Decision 3742/2001/QD-BYT[29] controls the use of additives in foodstuffs. The Decision lists permitted additives, which may be used in certain foodstuffs to the maximum levels stated.

5.3.5 MERCOSUR

Cristina Losada, Senior Regulatory Advisor, Leatherhead Food Research
Vanessa Richardson, Principal Regulatory Advisor, Leatherhead Food Research

Argentina, Brazil, Uruguay and Paraguay signed the Treaty for the Organisation of a Southern Common Market (MERCOSUR) in 1991, also known as the Asunción Treaty. Venezuela signed a membership agreement with MERCOSUR on 17 June 2006, but only became a full member on 31 July of 2012. A transitional period of four years is in place during which Venezuela will implement MERCOSUR Regulations.[30]

The aim of MERCOSUR is to create a free trade area and a customs union, and to issue legislation to avoid distorting operation of a common market. To facilitate trade, harmonized food legislation is being developed by the Common Market Group (GMC in Portuguese/Spanish). Each Member State must comply with the provisions of the Asunción Treaty and implement harmonized MERCOSUR legislation.

A general positive list of food additives was first established in 1993. However, in 2006, it was revoked and replaced by a new harmonised general list of additives (LGHA, in Portuguese/Spanish), which includes colours and sweeteners permitted throughout MERCOSUR (MERCOSUR GMC Resolution No. 11/06 of 22 July 2006[31]). The LGHA does not imply that all additives listed can be used in general. Permissibility of food additives in specific foodstuffs is normally established in commodity-specific legislation.

In 1996, MERCOSUR published a list of additives permitted according to Good Manufacturing Practice (GMP) that was revoked and replaced by MERCOSUR GMC Resolutions Nos. 34/10[32] and 35/10[33] of 15 July 2010. MERCOSUR Resolution GMC No. 34/10 is the revised list of GMP additives whereas No. 35/10 contains a list of food additives that were excluded from the previous GMP additives Resolution, along with their newly established conditions of use. Similarly to the LGHA, MERCOSUR Resolution GMC No. 34/10 does not imply that all GMP additives listed can be used in foodstuffs in general, as the acceptability of use of GMP additives is normally laid down in commodity-specific legislation.

MERCOSUR Resolutions must be implemented into Member States to be valid and therefore national law must be consulted to determine whether an additive is permitted or not. MERCOSUR-established Technical Regulations on the use of food additives have been laid down for a variety of commodities, such as meat and meat products,[34] confectionery[35] and bakery products.[36]

In addition to MERCOSUR-based legislation, some Member States also rely heavily on national provisions. For example, Brazil has several pieces of legislation, some of which go back as far as the 1950s, which are still in force. Similarly, the Argentinian Food Code contains many national provisions that are still applicable.

5.3.6   Middle East

Linda Amirat, Regulatory Advisor, Leatherhead Food Research
Ella Freeman, Senior Regulatory Advisor, Leatherhead Food Research

Key legislation in the context of the Middle East includes that of Saudi Arabia and the Gulf States. The Gulf Co-operation Council (GCC) is constituted by

Saudi Arabia, United Arab Emirates, Bahrain, Kuwait, Oman, Qatar and Yemen and follows Gulf Food Standards.

Gulf Standards are adopted by the Council of the GCC countries. In GCC, food additives are regulated by the following Gulf standards:

- GS 356/1994 on preservatives permitted for use in food products;
- GS 357/1994 on antioxidants permitted for use in food products;
- GS 381/1994 emulsifiers, stabilisers and thickeners permitted for use in food products;
- GS 23/1998 on colouring matters used in foodstuffs;
- GSO 995/1998 on sweeteners permitted for use in food products.

Generally, these standards provide only positive lists of additives. However, these must be permitted by the relevant Gulf product-specific standards in order to able to be used in the respective foodstuff. These product specific-standards also specify maximum levels for use of food additives.

Commodity food standards contain relevant provisions concerning specific foods. Gulf standards on salts of sulfurous acids used in preservation of foodstuffs (GS 175/1994) and the standard on benzoic acid, sodium benzoate and potassium benzoate used in preservation of foodstuffs (GS 172/1994) states the amount of benzoic acid allowed in foodstuffs. The Gulf Standard on sweeteners (GSO 995/1998) concerns sweeteners permitted in foods for particular nutritional use, such as diabetic food, or in energy-reduced foods or foods with no added sugars.

In the absence of specific authorisation for use of an additive in a food standard it is necessary to contact the authorities to obtain approval for its use. This can cause problems for manufacturers of products that are not standardised as individual approvals are required, which can be time consuming.

Import of alcohol (including foods prepared or preserved with alcohol) and cyclamates is not permitted. It is, therefore, advisable to use alternative solvents to alcohol-based compounds when exporting to the Middle East. Foodstuffs containing alcohol are not permitted in GCC countries. However, a threshold in the end product of less than 0.05% in the case of alcohol resulting from a process of fermentation may be accepted. This is again subject to confirmation with the authorities. Halal requirements must be taken into consideration when dealing with GCC countries. Food products containing gelatin of animal origin other than pork (as gelatin from pork is prohibited in GCC) must be accompanied by a halal certificate to justify that the animal was slaughtered according to the Muslim ritual.

Additives legislation in Israel is based upon European Parliament and Council Directive 94/36/EC on colours for use in foodstuffs,[37] European Parliament and Council Directive 94/35/EC, as amended, on sweeteners for use in foodstuffs,[38] and European Parliament and Council Directive 95/2/EC, as amended, on food additives other than colours and sweeteners,[39] implemented as The Public Health (Food Additives) Regulations 2001.[40] The annex to the Regulation lists all food additives permitted for use in the various foodstuffs. The majority of provisions on the use of additives and maximum levels laid down in the EU Directives are maintained in the Israeli Regulations, including the principles of *quantum satis* and carry-over, with some alterations or amendments made to a few specified additives or categories of foodstuffs in which the additives are permitted for use.

The provisions of Israeli compositional standards relating to the use of additives are superseded by the additives regulations, except for provisions on the use of additives in 13 foodstuffs typical for Israeli consumption, namely sesame *halvah*, soft white cheese, fermented milk products, grape juice, sesame *tehina*, pickled black olives, preserved grape juice, specific salads, mixed spices, salty cheeses and *matzoth* for Passover. The Regulations and annex are available in Hebrew only.

The Regulations are based upon the English version of the EU Directives and translated into Hebrew, therefore translation back into English may result in errors of interpretation. It is recommended that the EU Directives themselves be used as guidance.

The use of additives in raw materials used for the professional and industrial manufacture of foodstuffs are not covered by the Public Health (Food Additives) Regulations and therefore Israeli standards apply.

An additional complication for additives in Israel is that certain additives may not be Kosher and this may give rise to problems; for example, the use of animal-derived additives in a dairy product.

5.3.7 Australia and New Zealand

Mariko Kubo, Principal Regulatory Advisor, Leatherhead Food Research

The use of food additives in both Australia and New Zealand is regulated by the Australia New Zealand Food Standards Code (Code)[41] and enforced in both countries under State and Territory food laws. The use of additives in specific foods is detailed in Standard 1.3.1.

Food Standards Australia New Zealand (FSANZ) is responsible for the development of, or variation to, food standards established in the Code. FSANZ is an independent statutory agency established by the Food Standards Australia New Zealand Act 1991.[42] Working within an integrated food regulatory system involving the governments of Australia and the New Zealand Government, it set food standards for the two countries. FSANZ is part of the Australian Government's Health and Ageing portfolio.[41]

In New Zealand, the New Zealand Food Safety Authority (NZFSA) is responsible for the implementation of the Australia New Zealand Food Standards Code that took full effect on 20 December 2002. Food sold in New Zealand must comply with the Code.

The standard development process involves an evaluation of the risk to public health of the proposed change to the Code and the impact of the regulatory measures on the food industry and on the country's international trading obligations. Then FSANZ draft a legal standard for public comment. There may be one or more periods of public consultation for each standard.

Finally, the draft standard is considered for approval by the FSANZ Board and, if the Ministerial Council does not request a review of the decision within 60 days, the standard is published as law in the FSANZ gazette.

Before the use of a new additive in a food can be approved, it is necessary to have answers to the following questions:

- Is the additive safe to eat at the requested level in a particular food?
- Are there good technological reasons for the use of the additive?
- Will consumers be clearly informed about its presence?

Only if satisfaction is reached on these points will a recommended maximum permitted level in a particular food be put forward, based on technological need and provided it is well within safety limits.

FSANZ allows additives to be used only if it can be demonstrated that no harmful effects are expected, following evaluation of data obtained by testing the additive. Food additive safety is based on the Acceptable Daily Intakes (ADIs). A review of individual food standards is currently underway to bring them up to date with the modern food industry. Dietary evaluation has been carried out to ensure that consumption is well within safe limits, even for those who consume large quantities of certain foods. It is considered that the revised standard is more flexible for the industry, allowing technological developments while maintaining product safety.

The Australian New Zealand Food Standards Code is unusual in that it contains Standard 1.3.3 on processing aids. In many countries, processing aids are not specifically regulated; under this Standard the specific compounds that may be used for processing aid functions are detailed. The Code defines a processing aid as a substance used in the processing of raw materials, foods or ingredients to fulfil a technological purpose relating to treatment or processing, but which does not perform a technological function in the final food. In contrast to the situation in most countries, the use of processing aids in foods is prohibited unless specific provision is given for that use. The processing aids categories covered by the Standard include generally permitted processing aids; catalysts; bleaching agents, washing and peeling agents; extraction solvents; enzymes; microbial nutrients and microbial nutrient adjuncts; and those processing aids used in packaged water and in water used as an ingredient in other foods.

Enzymes of microbial, animal or plant origin are detailed, together with their authorised sources. For example, Bromelain from pineapple stem (*Ananascomosus*) and alpha-Galactosidase from *Aspergillus niger* are listed.

5.3.7.1 Labelling

Standard 1.2.4 of the Code covers provision on declaration of additives. Additives in the ingredients list must be declared using one of the 19 prescribed class names followed by the prescribed name or the code number, in brackets, of the additive as specified in the Food Standards Code. In addition to the 19 prescribed class names, six options class names such as antifoaming agent and modified starch are also provided in the Code. The code numbers are based on the International Numbering System. As in the European Union, it is the function of the additive in the food that determines additive class declaration; if an additive can perform more than one function, then the one most appropriate to the function being carried out in the particular food must be declared.

References

1. CAC, 2006. *Understanding the Codex Alimentarius* [Online]. 3rd edn. Rome: FAO. Available: ftp://ftp.fao.org/codex/Publications/understanding/Understanding_EN.pdf [Accessed 26 Jan, 2012].
2. CAC, 2011. *ProceduralManual* [Online]. 20th edn. Rome: FAO/WHO, Available: ftp://ftp.fao.org/codex/Publications/ProcManuals/Manual_20e.pdf [Accessed 26 Jan, 2012].
3. Codex Alimentarius Executive Committee, 2005. *Report of the fifty-fifth session of the Executive Committee of the Codex Alimentarius Commission, 9–11 February 2005* [Online]. Rome: FAO, Available: http://www.codexalimentarius.net/download/report/629/al28_03e.pdf [Accessed 26 Jan, 2012].
4. CAC, 1995. *GeneralStandardforFoodAdditives (CODEXSTAN192-1995), last revised in 2011* [Online]. Rome: FAO/WHO, Available: http://www.codexalimentarius.net/download/standards/4/cxs_192e.pdf [Accessed 26 Jan, 2012].
5. CAC, 1989. *Class Names and the International Numbering System for Food Additives (CAC/GL 36-1989), last amended in 2011* [Online]. Rome: FAO/WHO, Available: http://www.codexalimentarius.net/download/standards/7/CXG_036e.pdf [Accessed 26 Jan, 2012].
6. JECFA. *JECFA Glossary of Terms* [Online]. Available: http://www.who.int/foodsafety/chem/jecfa/glossary.pdf [Accessed 26 Jan, 2012].
7. FAO/WHO, 1973. *Toxicological Evaluation of Certain Food Additives with a Review of General Principles and of Specifications, 25 June–4 July 1973* [Online]. Geneva: WHO, Available: http://whqlibdoc.who.int/trs/WHO_TRS_539.pdf [Accessed 26 Jan, 2012].
8. JECFA, 2011. *Guidelines for the preparation of working papers on intake of food additives for the Joint FAO/WHO Expert Committee on Food Additives* [Online]. Geneva: WHO, Available: http://www.who.int/foodsafety/chem/jecfa/en/intake_guidelines.pdf [Accessed 26 Jan, 2012].

9. CAC, 2011. *List of Codex Specifications for Food Additives (CAC/MISC 6-2011),* [Online]. Rome: FAO/WHO, Available: http://www. codexalimentarius.net/download/standards/9/CXA_006e.pdf [Accessed 26 Jan, 2012].

10. USA. *Federal Food, Drug and Cosmetic Act (as amended)* [Online]. Available: http://www.fda.gov/RegulatoryInformation/Legislation/Federal FoodDrugandCosmeticActFDCAct/FDCActChapterIVFood/default.htm [Accessed 30 Jan, 2012].

11. USA. *Title 21 of the Code of Federal Regulations, Chapter 1, Parts 1-199* [Online]. Available: http://www.gpo.gov/fdsys/browse/collectionCfr.action? collectionCode=CFR&searchPath=Title + 21%2FChapter + I&oldPath= Title + 21&isCollapsed=true&selectedYearFrom=2011&ycord=474 [Accessed 30 Jan, 2012].

12. USA. *FDA. Food Additive Petition Expedited Review–Guidance for Industry* [Online]. Available: http://www.fda.gov/Food/GuidanceCompliance RegulatoryInformation/GuidanceDocuments/FoodIngredientsandPackaging/ ucm224104.htm [Accessed 30 Jan, 2012].

13. USA. *FDAFood* [Online]. Available: http://www.fda.gov/Food/default. htm [Accessed 30 Jan, 2012].

14. USA, 1997. *FDA Substances Generally Recognized as Safe; Proposed Rule. Federal Register,* 62, (74), 17/4/97, 18937–64 [Online] Available: http:// www.fda.gov/Food/FoodIngredientsPackaging/GenerallyRecognizedasSafe GRAS/ucm083058.htm [Accessed 30 Jan, 2012].

15. USA. *FDA GRAS Notice Inventory* [Online]. Available: http://www.accessdata. fda.gov/scripts/fcn/fcnNavigation.cfm?rpt=grasListing [Accessed 30 Jan, 2012].

16. EU, 1997. Regulation (EC) No. 258/97 of the European Parliament and of the Council of 27 January 1997 concerning novel foods and novel food ingredients, *Off. J. Eur. Communities,* 14/02/1997, **40**(L 43), 1–6.

17. USA. *Flavor and Extract Manufacturers Association (FEMA) FEMA GRAS™ List* [Online]. Website: http://www.femaflavor.org/fema-gras% E2%84%A2-flavoring-substance-list [Accessed 30 Jan. 2012].

18. CANADA, 1985. *Food and Drugs Act, as amended.* [Online] Available: http://laws-lois.justice.gc.ca/eng/acts/F-27/index.html [Accessed 30 Jan. 2012].

19. CANADA. *Food and Drug Regulations, as amended.* [Online] Available: http://laws-lois.justice.gc.ca/eng/regulations/C.R.C.%2C_c._870/ [Accessed 30 Jan. 2012].

20. CANADA. *Health Canada, Interim Marketing Authorisations* [Online]. Available: http://www.hc-sc.gc.ca/fn-an/legislation/ima-amp/index-eng. php [Accessed 30 Jan. 2012].

21. JAPAN, 1948. Ordinance for Enforcement of the Food Sanitation Law (Ministry of Health, Labour and Welfare (MHLW) Ordinance No. 23 of 13 July 1948), as amended.

22. JAPAN, 1959. Specifications and Standards for Food and Food Additives (MHLW Notification No. 370 of 28 December 1959), as amended.

23. JAPAN, 1996. List of Existing Food Additive (MHLW Notification No. 120 of 16 April 1996), as amended.
24. JAPAN, 2010. Additive Declaration in accordance with the Food Sanitation Law (Consumer Affair Agency Notice No. 377 of 20 October 2010), as last amended by Notice No. 246 of 28 June 2011.
25. MALAYSIA, 1985. Food Regulations 1985, as amended.
26. SINGAPORE, 1988. *Food Regulations 1988, as amended* [Online]. Available: http://www.ava.gov.sg/NR/rdonlyres/0CA18578-7610-4917-BB67-C7DF4B96504B/19280/2web_SOF_FoodRegulations15April2011.pdf [Accessed 30 Jan. 2012].
27. CHINA, 2011. National Food Safety Standard GB 2760-2011 Standard for the Use of Food Additives, as amended.
28. TAIWAN, 2008. *The Scope and Application Standards of Food Additives as amended* [Online]. Available:http://www.fda.gov.tw/eng/people_laws_list.aspx?time=1&classifysn=16 [Accessed 30 Jan, 2012].
29. VIETNAM, 2001. The Ministry of Health of Vietnam Decision 3742/2001/QD-BYT.
30. MERCOSUR. *About MERCOSUR* [Online]. Available: http://www.mercosur.int/t_generic.jsp?contentid=661&site=1&channel=secretaria&seccion=2 [Accessed 27 Jan, 2012].
31. MERCOSUR, 2006. *MERCOSUR/GMC/RES No. 11/06. Technical Regulation on "Harmonised General List of Food Additives and their Functional Classes"* [Online]. Buenos Aires: MERCOSUR, Available: http://gd.mercosur.int/SAM%5CGestDoc%5Cpubweb.nsf/694D523EFFB0FDEF03257992003D8E5D/$File/RES_011-2006_PT_RTMAditivosAlimentares.pdf [Accessed 27 Jan, 2012].
32. MERCOSUR, 2010. *MERCOSUR/GMC/RES No. 34/10. MERCOSUR Technical Regulation on Food Additives Authorised for Use According to Good Manufacturing Practice (GMP)* [Online]. Buenos Aires: MERCOSUR, Available: http://gd.mercosur.int/SAM%5CGestDoc%5Cpubweb.nsf/E92613F46B0C4F4603257992003E4DFF/$File/RES_034-2010_PT_%20Aditivos%20BPF.pdf [Accessed 27 Jan, 2012].
33. MERCOSUR, 2010. *MERCOSUR/GMC/RES No. 35/10. MERCOSUR Technical Regulation on Maximum Limits for Additives Excluded from the List of "Food Additives Authorised for Use According to Good Manufacturing Practice"* [Online]. Buenos Aires: MERCOSUR, Available: http://gd.mercosur.int/SAM/GestDoc/PubWeb.nsf/OpenFile?OpenAgent&base = SAM\GestDoc\DocOfic0.nsf&id=5FEA70D72B3853BA8325774900543649&archivo=RES_035-2010_PT_ExcluidosListaAditivosBPF.pdf [Accessed 27 Jan, 2012].
34. MERCOSUR, 1997. *MERCOSUR/GMC/RES No. 73/97. MERCOSUR Technical Regulation on the Use of Food Additives and their Limits in the Following Food Categories: Category: Meat and Meat Products* [Online]. Montevideu: MERCOSUR, Available: http://gd.mercosur.int/SAM%5CGestDoc%5Cpubweb.nsf/D46521F3A470C1B90325799200412A16/$File/RES_073-1997_ES_RTMAditiv8.pdf [Accessed 27 Jan, 2012].

35. MERCOSUR, 1998. *MERCOSUR/GMC/RES No. 53/98. Technical Regulation on the "Use of Food Additives and their Maximum Limits for Food Category 5: Sweets, Comfits, Bonbons, Chocolates and Similar Products* [Online]. Rio de Janeiro: MERCOSUR, Available: http://gd.mercosur.int/SAM%5CGestDoc%5Cpubweb.nsf/6090C51422A D683E0325799200419326/$File/RES_053-1998_PT_RTM%20Atrib%20 Adit%20Categ%20Alimentos%205%20Similares_Ata%20%204_98.pdf [Accessed 27 Jan, 2012].

36. MERCOSUR, 1997. *MERCOSUR/GMC/RES No. 50/97. Technical Regulation on the Use of Food Additives and their Maximum Concentration For Food Category 7: Bakery Products* [Online]. Montevideu: MERCOSUR, Available: http://gd.mercosur.int/SAM%5CGestDoc% 5Cpubweb.nsf/479274A15231B944032579920041EC20/$File/RES_050-1997_ES_AsignAditAlim.pdf [Accessed 27 Jan, 2012].

37. EC, 1994. European Parliament and Council Directive 94/36/EC of 30 June 1994 on colours for use in foodstuffs, *Official Journal of the European Communities,* 10/09/1994, **37**(L 237), 13–29.

38. EC, 1994. European Parliament and Council Directive 94/35/EC of 30 June 1994 on sweeteners for use in foodstuffs, *Official Journal of the European Communities,* 10/09/1994, **37**(L 237), 3–12.

39. EC, 1995. European Parliament and Council Directive 95/2/EC of 30 June 1994 on food additives other than colours and sweeteners, *Official Journal of the European Communities,* 18/03/1995, **38**(L 61), 1–40.

40. The Public Health (Food Additives) Regulations 2001, as amended [Online]. Available: http://www.old.health.gov.il/Download/pages/fcs_ nov2010.pdf.pdf [Accessed 1 Feb, 2012].

41. AUSTRALIA and NEW ZEALAND, 2002. *Australia New Zealand Food Standards Code* [Online]. Available: http://www.foodstandards.gov.au/ foodstandards/foodstandardscode.cfm [Accessed 30 Jan, 2012].

42. AUSTRALIA and NEW ZEALAND, 1991.*Food Standards Australia New Zealand Act 1991*[Online]. Available: http://www.acfs.go.th/km/ download/FoodStandands_A_NZ_1991.pdf [Accessed 30 Jan, 2012].

CHAPTER 6
E Numbers

Number	Product
E100	Curcumin
	CI natural yellow 3
	Turmeric yellow
	Diferoyl methane
	Colour Index No: 75300

Sources
Curcumin is the principle pigment of turmeric, a spice that is obtained from the rhizomes of *Curcuma longa*, a perennial plant of the ginger family native to South Asia, which has been consumed for many thousands of years. Curcumin is obtained by solvent extraction from the rhizomes to produce an oleo-resin, which is then purified by crystallisation.
Curcumin is also available as an aluminium lake.

Function in Food
Curcumin is a natural colour that provides a bright lemon-yellow colour when used in foods. Although oil soluble, it is available in water-dispersible forms.
The pure pigment has a high tinctorial strength with an absorption maximum in the region of 426 nm when measured in acetone.

Benefits
The stability of curcumin to heat is excellent and it may generally be used in products throughout the acid pH range. Typical dose applications calculated on the basis of the pure pigment are between 10 and 100 ppm.

Limitations
Curcumin has poor stability to light and is sensitive to sulfur dioxide at levels in excess of 100 ppm. It is included in Part C Group III in EU Regulation

Essential Guide to Food Additives, 4th Edition
Edited by Mike Saltmarsh
© The Royal Society of Chemistry 2013
Published by the Royal Society of Chemistry, www.rsc.org

1129/2011, food colours with combined maximum limit. It has an ADI of 0–3 mg/kg body weight allocated by JEFCA. In the USA turmeric oleoresin is exempt from certification and is permanently listed for food use according to GMP.

From 1 August 2014 the aluminium lake is only permitted in food categories where a maximum limit on aluminium from lakes is explicitly stated in Part E of Regulation 1129/2011.

Typical Products
Smoked white fish, ice creams, sorbets, dairy products and some types of sugar confectionery.

Number	Product
E101	Riboflavin
	(i)Lactoflavin
	Vitamin B_2
	Riboflavin-5′-phosphate
	(ii)Riboflavin-5′-phosphate, sodium

Sources
Riboflavin is found in green vegetables, milk, eggs and yeast. It is produced commercially as yellow crystals by chemical synthesis. Riboflavin-5′-phosphate sodium is the more water-soluble form.

Function in Food
Riboflavins are used to provide a bright lemon yellow, but the colour can be incidental to the use as a vitamin.

Benefits
Both forms are nutritionally important as vitamin B2 and the colouring properties are often of secondary importance. They have good stability to heat and moderate stability to acid pH.

Limitations
Stability to light when in solution is poor and so applications in aqueous solution should be protected from light. Riboflavin is only slightly soluble in water and riboflavin – 5′-phosphate should be used where solubility is an important factor. Both products can impart a bitter taste. JEFCA has allocated ADI of 0–0.5 mg/kg body weight. Within the EU, they are listed in Group II of Regulation 1129/2011, food colours authorised at *quantum satis*. In the USA Riboflavin is exempt from certification and listed for food uses according to GMP.

Typical Products
Salad dressings, confectionery, tablet coatings and powdered drinks.

Number	Product
E102	Tartrazine
	CI food yellow 4
	FD&C yellow no.5

Colour Index No: 19140

Sources
Tartrazine is a water-soluble synthetic dye commercially available as the sodium salt of the dye. It is also available as the aluminium lake, which is water insoluble. Tartrazine is classed as a monoazo dye.

Function in Food
Tartrazine is a synthetic colour that is used to provide a bright, typically lemon-yellow colour to foods. The colour is lemon yellow but less green than quinoline yellow. It can be used in combination with other colours.
The pure pigment has a high tinctorial strength with an absorption maximum in water of 426 nm.

Benefits
Tartrazine has good stability to heat and light and is permitted in many countries.

Limitations
In general use, tartrazine shows some instability to ascorbic acid. The water solubility is very dependent on temperature (from 200 g/l at 25 °C to less than a quarter of that in cold conditions) and so it can come out of solution when cold but will easily redissolve when the mixture is warmed. It has been allocated an ADI of 0–7.5 mg/kg body weight by JECFA (1966) and EFSA (2009). It is included in Part C Group III of Regulation 1129/2011 (food colours with combined maximum limit) and is permitted in the USA as FD&C Yellow no.5.
From 1 August 2014 the aluminium lake is only permitted in food categories where a maximum limit on aluminium from lakes is explicitly stated in Part E of Regulation 1129/2011.
When tartrazine is used in a foodstuff in the EU the labelling must include the following phrase as required by article 24 and Annex V of Regulation 1333/2008 "Tartrazine (or E102) may have an adverse effect on activity and attention in children".

Typical Products
Confectionery, pickles, sauces and seasonings.

Number   Product
E104   Quinoline yellow
                                          CI food yellow 13

                              Colour Index No: 47005

Sources
Quinoline yellow is a water-soluble dye commercially available as the sodium
salt of the dye. It is also available as the aluminium lake, which is water
insoluble. Quinoline yellow is classed as a quinophthalone dye.
The commercial material consists essentially of a mixture of disulfonates and
monosulfonates with the disulfonates in the majority. Another dye, D&C
yellow no. 10, consists mainly of monosulfonates but shares the same Colour
Index number.

Function in Food
Quinoline yellow is a synthetic colour that is used to provide a greenish-yellow,
simulating the shade of pineapple and lemon. It is often used in combination
with other colours.
The pure pigment has a high tinctorial strength with an absorption maximum
in water of 411 nm.

Benefits
It has good stability to heat and light and is generally stable in the presence of
fruit acids and sulfur dioxide.

Limitations
Quinoline yellow is generally stable in foods. It is included in EU Regulation
1129/2011 but the permitted levels are very restricted. It has been allocated an
ADI of 0–0.5 mg/kg body weight by EFSA (2009) and a temporary ADI of
0–5 mg/kg by JECFA (2011).
Quinoline yellow is not permitted in foods in the USA; D&C Yellow No 10 is
only permitted in drugs and cosmetics.
From 1 August 2014 the aluminium lake is only permitted in food categories
where a maximum limit on aluminium from lakes is explicitly stated in Part E
of Regulation 1129/2011.
When quinoline yellow is used in a foodstuff in the EU, the labelling must
include the following phrase as required by article 24 and Annex V of Regulation 1333/2008 "Quinoline yellow (or E104) may have an adverse effect on
activity and attention in children".

Typical Products
Soft drinks, confectionery, pickles, sauces and seasonings.

Number	Product
E 110	Sunset yellow FCF
	CI food yellow 3
	Orange yellow S
	FD&C yellow no.6

Colour Index No: 15985

Sources
Sunset yellow is a water-soluble synthetic dye commercially available as the sodium salt of the dye. It is also available as the aluminium lake, which is water insoluble. Sunset yellow is classed as a monoazo dye.

Function in Food
Sunset yellow is a synthetic colour that is used to provide an orange shade characteristic of orange peel. It is often used in combination with other colours. The pure pigment has a high tinctorial strength with an absorption maximum in water of 485 nm.

Benefits
Sunset yellow has good stability to heat and light.

Limitations
Although it is generally stable in solution between pH levels of 3 and 8, sunset yellow is only moderately stable in the presence of benzoic acid and sulfur dioxide, fading in the latter. The presence of calcium ions can cause the colour to precipitate.

In the EU it is included in Regulation 1129/2011 Annex II, but with very strict limits due to the low temporary ADI. It is certified by the FDA in the USA as FD&C yellow no.6. It has been allocated a temporary ADI of 0–1 mg/kg body weight by EFSA (2009) and 0–4 mg/kg body weight by JECFA (2011).

After 1 August 2014 the aluminium lake is only permitted in food categories where a maximum limit on aluminium from lakes is explicitly stated in Part E of Regulation 1129/2011.

When sunset yellow is used in a foodstuff in the EU, the labelling must include the following phrase as required by article 24 and Annex V of Regulation 1333/2008 "Sunset yellow (or E110) may have an adverse effect on activity and attention in children".

Typical Products
Soft drinks

Number	Product
E 120	Cochineal
	Carminic acid
	Carmine

Colour Index No: 75470

Sources

Carminic acid is the red pigment obtained by aqueous alkaline extraction from the dried bodies of the coccid insect (*coccus cacti L.*) which lives on cactus plants in South America and Mexico. Cochineal is properly the term for the dried bodies of the insects but is commonly used for both the insects and the colour extracted from them. Carmine pigment is the aluminium lake of carminic acid.

Function in Food

Carmine has a long history of use as a food colour that provides a bright strawberry red shade to a wide variety of products. It is generally used in products in which the pH is above 3.5 and is available in both water-insoluble and water-soluble forms.

Carminic acid is water soluble and is particularly appropriate for providing clear orange hues in acid-based preparation such as soft drinks.

The absorption maximum for carmine is in the region of 518 nm when measured in aqueous ammonia solution. Carminic acid has an absorption maximum in the region of 494 nm in dilute hydrochloric acid.

Benefits

Both carminic and carmine are chemically very stable, with excellent resistance to oxygen, light, sulfur dioxide, heat and water activity. Typical dose applications calculated on the basis of the pure pigment are between 5 and 50 ppm.

Limitations

Carmine precipitates in low pH conditions and should not be used in foods in which the pH is below 3.5. The shade of carminic acid is dependent upon pH, and it is generally suitable only for the colouration of acidic products.

Carmine and carminic acid are both allocated an ADI of 0–5 mg/kg body weight and are included in Group III of Part C of EU Regulation 1129/20011, food colours with a combined maximum limit. Carmine is permitted in the USA in accordance with GMP.

JECFA evaluated potential allergenicity and concluded that cochineal extract, carmines, and possibly carminic acid in foods and beverages may provoke allergic reactions in some individuals.

From 1 August 2014 the aluminium lake is only permitted in food categories where a maximum limit on aluminium from lakes is explicitly stated in Part E of Regulation 1129/2011.

Neither carmine nor carminic acid may be used in products claiming their suitability for vegetarian diets.

Typical Products
Meat products, beverages, table jellies, sugar confectionery, yogurts and desserts.

Number	Product
E 122	Carmoisine
	Azorubine
	CI food red 3
	Colour Index No. 14720

Sources
Carmoisine is a water-soluble synthetic dye commercially available as the sodium salt of the dye. It is also available as the aluminium lake, which is water insoluble. Carmoisine is classed as a monoazo dye.

Function in Food
Carmoisine is a synthetic dye used to provide a bluish red colour appropriate for raspberry or blackcurrant flavoured foods.
The pure pigment has a high tinctorial strength with an absorption maximum in water of 515 nm.

Benefits
Carmoisine is an intense raspberry red shade and can be used to produce purple shades when mixed with blue shades.

Limitations
Carmoisine is more susceptible to reducing agents than the other azo food colours and can fade in the presence of ascorbic acid. In the EU, it is included in Regulation 1129/2011 Part C Group III (food colours with combined maximum limit). It has been allocated an ADI of 0–4 mg/kg body weight by EFSA (2009) and JECFA (1983).
After 1 August 2014 the aluminium lake is only permitted in food categories where a maximum limit on aluminium from lakes is explicitly stated in Part E of Regulation 1129/2011.
When carmoisine is used in a foodstuff in the EU, the labelling must include the following phrase as required by article 24 and Annex V of Regulation 1333/2008 "Carmoisine (or E122) may have an adverse effect on activity and attention in children".

Typical Products
Soft drinks, desserts and confectionery

Number	Product
E 123	Amaranth
	CI food red 9
	FD&C red no.2

Colour Index No: 16185

Sources
Amaranth is a water-soluble synthetic dye commercially available as the sodium salt of the dye. It is also available as the aluminium lake, which is water insoluble. Amaranth is classed as a monoazo dye.

Function in Food
Amaranth is a synthetic dye used to provide a deep bluish red shade typical of red berry foods. The shade is slightly bluer than Carmoisine. The pure pigment has a high tinctorial strength with an absorption maximum in water of 520 nm.

Limitations
Within the EU, amaranth is only permitted in Regulation 1129/2011 in fish roe, spirit drinks and wine product cocktails. Amaranth is not authorised for use in the USA but is permitted in Canada and South America.
It has been allocated an ADI of 0–0.015 mg/kg body weight by EFSA (2009) and 0–0.5 mg/kg bw by JECFA (1984)
After 1 August 2014 the aluminium lake is only permitted in food categories where a maximum limit on aluminium from lakes is explicitly stated in Part E of Regulation 1129/2011.

Typical Products
Fish roe and alcoholic beverages

Number	Product
E 124	Ponceau 4R
	Cochineal red A
	CI food red 7
	New coccine

Colour Index No: 16255

Sources
Ponceau 4R is a water-soluble synthetic dye commercially available as the sodium salt of the dye. It is also available as the aluminium lake, which is water insoluble. Ponceau 4R is classed as a monoazo dye.

Function in Food
Ponceau 4R is a synthetic dye used to provide a bright red shade typical of strawberry, cherry, or redcurrant-flavoured foods. It is used in combination

with other colours to provide the desired hue. The pure pigment has a high tinctorial strength with an absorption maximum in water of 505 nm.

Benefits
Ponceau 4R is a particularly bright red colour which has good light and heat stability.

Limitations
Within the EU it is included in Regulation 1333/2008 Annex II with very strict limits due to its low ADI. It has been allocated an ADI of 0–0.7 mg/kg body weight by EFSA (2009). JECFA allocated an ADI of 0–4 mg/kg body weight (2011). It is not permitted in food in the USA.
After 1 August 2014 the aluminium lake is only permitted in food categories where a maximum limit on aluminium from lakes is explicitly stated in Part E of Regulation 1129/2011.
When Ponceau 4R is used in a foodstuff in the EU, the labelling must include the following phrase as required by article 24 and Annex V of Regulation 1333/2008 "Ponceau 4R (or E124) may have an adverse effect on activity and attention in children".

Typical Products
Soft drinks, edible ices, confectionery and desserts.

Number	Product
E 127	Erythrosine
	FD&C red no. 3
	Colour Index No: 45430

Sources
Erythrosine is a water-soluble synthetic dye commercially available as the sodium salt of the dye. It is also available as the aluminium lake, which is water insoluble. It is classed as a xanthene dye and is manufactured by treatment of a boiling alcoholic solution of fluorescein with iodine and sodium iodide.

Function in Food
Erythrosine is a synthetic dye used to impart a light red colour to foods. The pure pigment has a high tinctorial strength with an absorption maximum in water of 526 nm.

Benefits
Erythrosine imparts a unique bright pink shade with good stability to heat and it is highly stable in the presence of sulfur dioxide. The dye is particularly suitable for colouring maraschino and canned cherries, where no bleeding is essential. It is precipitated in the fruit by the fruit acids naturally present.

Limitations
In acidic sugar confectionery erythrosine undergoes a colour change to an orange shade. Stability to light is poor and it is precipitated in acidic conditions to its free acid. The erythrosine molecule contains iodine, which has been linked to thyrotoxicosis, as a result of which the use of the dye has been restricted in its applications. It has been allocated an ADI of 0–0.1 mg/kg body weight by EFSA (2011) and JECFA (1990).
After 1 August 2014 the aluminium lake is only permitted in food categories where a maximum limit on aluminium from lakes is explicitly stated in Part E of Regulation 1129/2011.
In the EU erythrosine is permitted in Regulation 1129/2011 for use only in cocktail cherries, candied cherries and bigareaux cherries in syrup.

Number	Product
E 129	Allura red AC
	CI food red 17
	FD&C red no. 40
	Colour Index No: 16035

Sources
Allura is a water-soluble synthetic dye commercially available as the sodium salt of the dye. It is also available as the aluminium lake, which is water insoluble. Allura red is classed as a monoazo dye.
Allura red was developed in the USA as FD&C red no. 40 to fill the gap arising from the delisting of FD&C red no. 4 and FD&C red no.2.

Function in Food
Allura red provides an orange red shade in solution, somewhat weaker in strength than other reds. In the USA it produces the characteristic shade of blueberries. The pure pigment has a medium tinctorial strength with an absorption maximum in water of 504 nm.

Benefits
Allura red is a general purpose colour with reasonable stability in a variety of foods and tolerance to processing and storage.

Limitations
Allura red is not very stable in the presence of oxidising and reducing agents and tends to go bluer in alkaline conditions. It does not provide a good dye for blending purposes; mixtures containing allura red tend to be dull. As a somewhat weak shade, it needs to be used at concentrations of around 100 mg/kg to produce an adequate coloration. Within the EU it is permitted in Group III of Part C of Regulation 1129/2011 (food colours with combined maximum limit). It has been allocated an ADI of 0–7 mg/kg body weight by

JEFCA. Allura red is permitted in the USA for uses in food and drugs according to GMP.

After 1 August 2014 the aluminium lake is only permitted in food categories where a maximum limit on aluminium from lakes is explicitly stated in Part E of Regulation 1129/2011.

When used in a foodstuff in the EU, the labelling must include the following phrase as required by article 24 and Annex V of Regulation 1333/2008 "Allura red (or E129) may have an adverse effect on activity and attention in children".

Typical Products
Soft drinks, confectionery, edible ices and desserts.

Number	Product
E 131	Patent blue V
	CI blue 5

Colour Index No: 42051

Sources
Patent blue V is a water-soluble synthetic dye commercially available as either the sodium or calcium salt of the dye. It is also available as the aluminium lake, which is water insoluble. Patent blue V is classed as a triarylmethane dye.

Function in Food
Patent blue V is a bright blue colour that is mainly used in combination with tartrazine or quinoline yellow to provide a green colour. The only significant use of the colour alone is in carcass staining.

Benefits
Patent blue V is highly stable to heat and light, but less stable to food ingredients such as fruit acids and benzoic acid.

Limitations
Although available in commercial quantities, this dye has a limited number of manufacturers. It should not be confused with the dye blue VRS, which has a very similar molecular structure, but that is not a permitted food dye.

Moderate solubility in water is a limitation but, with a maximum of 4–6% at room temperature, is sufficient for use at levels of 50–200 mg/kg of finished product.

With in the EU, Patent blue V is permitted in Regulation 1129/2011 Part C Group III (food colours with combined maximum limit). It has been allocated an ADI of 0–15 mg/kg (EFSA 1983).

After 1 August 2014 the aluminium lake is only permitted in food categories where a maximum limit on aluminium from lakes is explicitly stated in Part E of Regulation 1129/2011.

Typical Products
Baked goods and confectionery.

Number	Product
E 132	Indigo carmine
	Indigotine
	CI food blue 1
	FD&C blue no.2

Colour Index No. 73015

Sources

Indigo carmine is a water-soluble synthetic dye commercially available as the sodium salt of the dye. It is also available as the aluminium lake, which is water insoluble. Indigo carmine is classed as an indigoid dye.

Function in Food

Indigo carmine is use to provide a dark bluish-red colour to foodstuffs. It is often used in combination with other colours.

The pure pigment has a high tinctorial strength with an absorption maximum in water of 610 nm.

Benefits

Indigo carmine has wide acceptability and it is often used in combination with other dyes.

Limitations

The major limitation of this dye is its instability. It is unstable to most conditions – heat, light and common food ingredients. It fades in the presence of sulfur dioxide, sugar and syrups. Fading increases with pH across the range pH5 to 8 and it fades completely at pH9. Its limited solubility in water at 1–2% also restricts its use but it can be used at concentrations of 50–200 ppm.

It is approved for use in food in the USA as FD&C blue No.2. In the EU, it is included in Regulation 1129/2011 Part C Group III (food colours with combined maximum limit). Indigo carmine has been allocated an ADI of 0–5 mg/kg body weight (EFSA 1983 and JECFA 1974).

After 1 August 2014 the aluminium lake is only permitted in food categories where a maximum limit on aluminium from lakes is explicitly stated in Part E of Regulation 1129/2011.

Typical Products

Confectionery, baked goods and edible ices.

Number	Product
E 133	Brilliant blue FCF
	CI food blue 2
	FD&C blue no. 1

Colour Index No: 42090

Sources

Brilliant blue FCF is a water-soluble synthetic dye commercially available as the sodium salt of the dye and it is also available as the water-insoluble aluminium lake. Brilliant blue is classed as a triarylmethane dye.

Function in Food
Brilliant blue provides a greenish blue colour, which is particularly used for blending with tartrazine or quinoline yellow to give greens and with other colours to give browns or blacks. The pure pigment has a high tinctorial strength with an absorption maximum in water of 630 nm.

Benefits
Brilliant blue is a very stable colour.

Limitations
Brilliant blue tends to fade at pH8 and above. It has been allocated an **ADI** of 6 mg/kg body weight by EFSA (2010) and 12.5 mg/kg bodyweight by JECFA (1970). It is permitted in the USA as FD&C blue no.1. Within the EU it is permitted in Regulation 1129/2011 Part C Group III (food colours with combined maximum limit).
After 1 August 2014 the aluminium lake is only permitted in food categories where a maximum limit on aluminium from lakes is explicitly stated in Part E of Regulation 1129/2011.

Typical Products
Soft drinks, confectionery, baked goods, desserts and edible ices.

Number	Product
E 140	(i) Chlorophyll CI natural green 3 Magnesium chlorophyll Magnesium phaeophytin Colour Index No: 75470
	(ii) Chlorophyllin CI natural green 5 Sodium chlorophyllin Potassium chlorphyllin Colour Index No: 75815

Sources
The oil-soluble chlorophylls are oil-soluble colours extracted from edible plant material including grass, lucerne (alfalfa) and nettle using one of a limited range of organic solvents or carbon dioxide. Alkaline saponification produces the chlorophyllins, which are soluble in water.

Function in Food
Chlorophylls and chlorophyllins are naturally derived colours that provide olive green hues to food products. Commercially, products are available for use in both oil- and water-based systems.

The absorption maximum of chlorophylls is in the region of 409 nm, measured in chloroform, and that of chlorophyllins around 405 nm, measured in aqueous pH9 buffer solution.

Benefits
Chlorophylls are natural pigments present in all green leafy vegetation, and they have always been a component of man's diet.

Limitations
Chlorophyll and chlorophyllins provide dull olive green hues and as a result have limited use in foodstuffs. Chlorophylls are less stable to light and acidic condition than their copper counterparts (E141). Usage rates are between 20 and 200 ppm.
They have been allocated an ADI of "not specified" by EFSA and are approved colours for use in Group II of Part C in EU Regulation 1129/2011, food colours authorised at *quantum satis*. They are not approved for use in the USA.

Typical Products
Sugar confectionery, yogurt and ice cream

Number   Product
E 141   (i) Copper complexes of Chlorophylls
                                                    CI natural green 3
                                                    Copper chlorophyll
                                                    Copper phaeophytin

                                               (ii) Copper complexes of Chlorophyllins
                                                    CI natural green 5
                                                    Sodium copper chlorophyllin
                                                    Potassium copper chlorophyllin
                                                    Colour Index No: 75815

Sources
Copper chlorophylls are oil-soluble colours obtained by addition of a copper salt to chlorophylls. Alkaline saponification prior to reaction with the copper salt produces the copper chlorophyllins, which are soluble in water.

Function in Food
Copper complexes of chlorophylls and chlorophyllins are chemically modified natural extracts that are used as colours providing blue-green hues to food products. Commercially, products are available for use in both lipid and aqueous media.
The absorption maximum of both products is in the region of 405–425 nm when measured, respectively, in chloroform or aqueous pH7.5 buffer solution.

Benefits
Copper chlorophylls provide brighter and more stable colours than the original chlorophylls. Typical dose applications are between 20 and 200 ppm calculated on the basis of the pure pigment.

Limitations
As chemically modified extracts, the copper chlorophylls should not be described as natural colours. The ADI allocated for both by EFSA is 0–15 mg/kg body weight.
Within the EU, copper chlorophylls are included in Group II of Part C in Regulation 1129/2011. They are not permitted as food colours in the USA.

Typical Products
Yogurts, ice cream, sauces, pickles.

Number	Product
E 142	Green S
	CI food green 4
	Brilliant green BS

Sources
Green S is a water-soluble synthetic dye commercially available as the sodium salt. It is also available as the aluminium lake, which is water insoluble. It is classed as a triarylmethane dye.

Function in Food
Green S delivers a greenish-blue colour in solution, characteristically more blue than its name suggests. It is most commonly used to produce green shades with tatrazine and quinoline yellow. It is traditionally the colour used for canned peas. The pure pigment has a high tinctorial strength with an absorption maximum in water of 632 nm.

Benefits
Green S is used in combination with other colours to produce brown and black shades.

Limitations
Green S has very few manufacturers as it has a complex structure and is difficult to synthesise. Within the EU, green S is permitted in Regulation 1129/2011 Part C Group III (food colours with combined maximum limit). It has been allocated an ADI of 5 mg/kg body weight by EFSA (2010).
After 1 August 2014 the aluminium lake is only permitted in food categories where a maximum limit on aluminium from lakes is explicitly stated in Part E of Regulation 1129/2011.

Typical Products
Canned peas and soft drinks.

Number	Product	
E150a	Class I	Plain caramel
E150b	Class II	Caustic sulfite caramel
E150c	Class III	Ammonia caramel
E150d	Class IV	Sulfite ammonia caramel

Sources

Caramel colours are dark brown to black liquids or powders that are manu-factured by the controlled heat treatment of carbohydrates.

Reactants may be used to promote the browning process. Different types of reactants are used to confer the appropriate physical characteristics to the caramel colour in correspondence to its final application.

There are 4 classes of caramel colours:

- Class I (E150a): Plain caramel is prepared with or without acids, bases or salts.
- Class II (E150b): Caustic sulfite caramel is prepared with or without acids or alkalis in the presence of sulfite compounds.
- Class III (E150c): Ammonia caramel is prepared with or without acids or alkalis in the presence of ammonium compounds.
- Class IV (E150d): Sulfite ammonia caramel is prepared with or without acids or alkalis in the presence of sulfite and ammonium compounds.

As the caramelization process can be exothermic, process control is critical to obtaining the correct colour.

Burnt sugars are light to dark brown liquids or powders that are obtained from controlled heat treatment of sugars. Burnt sugars can be used for both functions; bringing a colour or a flavour to the final product. Broadly, burnt sugars used for flavouring are considered ingredients whereas those used as colouring are additives and would be declared on the label as plain caramel. For detailed discussion of this subject, consult the decision tree on the website of the European Technical Caramel Association (EUTECA) http://www.euteca.org/PDF/EUTECA_decision-tree_on_e150a_ burnt_sugar.pdf.

Function in Food

Caramel colours are used to add or restore colour in food: the colour produced can be from brown red or yellow through dark brown to almost black.

They are typically used for colouring drinks, both alcoholic and nonalcoholic, and in baking applications, sauces, soups and gravies.

Benefits

The different production processes result in caramels with different properties. In particular, the different ionic charges of the different caramels provide different compatibilities.

Ammonia caramels are strongly positively charged. The positive ionic charge makes these caramels compatible with products containing proteins. A typical application consists in colouring beer, sauces, dairy products or pet food.

The sulfite ammonia caramels are negatively charged, so they are more compatible with the acidic environment of soft drinks. This class is the most used because the major application of caramel colour is in cola drinks.

Plain caramels are neutral to slightly negative. They are stable in high concentrations of alcohol. That is why one of their main application is in colouring spirits.

Caramel colours are totally miscible with water.

Limitations

Within the EU, the use of caramel colours is controlled by Regulation 1129/2011 where they are included in Group II, food colours authorised at *quantum satis*.

There are no technical limitations on the use of caramels, although it is important to select the type that is most appropriate for the intended use.

The process of reevaluation of caramel colours is underway; EFSA produced its Scientific Opinion in March 2011 and the European Commission (DG SANCO) is in the process of producing its findings on the basis of the positive EFSA assessment.

Typical Products

Soft drinks, beer, gravies and sauces, meat products.

Number	Product
E 151	Black PN
	Brilliant black BN
	CI food black 1
	Colour Index No: 28440

Sources

Black PN is a water-soluble synthetic dye commercially available as the sodium salt of the dye. It is also available as the aluminium lake, which is water insoluble. Black PN is classed as a bisazo dye.

Function in Food

Black PN is a purple to blue-black colour on its own but is mainly used for blending to provide violet to purple shades. The pure pigment has a high tinctorial strength with an absorption maximum in water of 570 nm.

Benefits

Black PN has a unique purple to blue-black shade.

Limitations
Although moderately stable to light, this dye has poor heat stability and, whilst stable in alkaline conditions, it is not very stable in the presence of fruit acids and sulfur dioxide. Black PN is allocated an ADI of 0–5 mg/kg bodyweight by EFSA (2010) and 1 mg/kg bodyweight by JECFA (1981). Within the EU it is included in 1129/2011 Part C Group III (food colours with a combined maximum limit).

After 1 August 2014 the aluminium lake is only permitted in food categories where a maximum limit on aluminium from lakes is explicitly stated in Part E of Regulation 1129/2011.

Typical Products
Fish roe products.

Number	Product
E 153	Vegetable carbon
	Vegetable black

Colour Index No: 77266

Sources
Vegetable carbon is manufactured by heating vegetable material to a high temperature in the absence of air. While wood, cellulose residue or coconut shell can be used, the main source is peat, as the product from this source tends to have the lowest ash content.

Function in Food
Vegetable carbon is a black powder, insoluble in water or organic solvents, which is used to darken the colour of solid foodstuffs.

Benefits
Vegetable carbon is inert, odourless and tasteless.

Limitations
There are only limited manufacturing sources for this product, which is difficult to handle and use as a powder. It is commonly prepared as a paste or dispersion before incorporation into a product. Although it is widely permitted, it is rarely used since low levels of use tend to produce grey shades, while the level necessary to produce black is often above rates that would be considered Good Manufacturing Practice.

Vegetable carbon has been allocated an ADI of "not specified" by EFSA. Within the EU, it is included in Part C Group II of Regulation 1129/2011, food colours authorised at *quantum satis*.

Typical Products
Confectionery, particularly liquorice.

Number	Product
E 155	Brown HT
	CI food brown 3
	Chocolate brown HT
	Colour Index No: 20285

Sources
Brown HT is a bisazo water-soluble synthetic dye. It is manufactured as the sodium salt and has been prepared as the water-insoluble aluminium lake.

Function in Food
A reddish-brown colour in solution, this dye has found most use in the baking industry – hence the suffix, HT which means "high temperature". The pure pigment has a high tinctorial strength with an absorption maximum in water of 460 nm.

Benefits
Brown HT is soluble in both water and propylene glycol. It is stable to heat and light and to the action of both alkalis and fruit acids.

Limitations
Brown HT has no technical limitations to its use. It has been allocated an ADI of 0–1.5 mg/kg body weight by EFSA (2010) and JECFA (1984), and in the EU is included in Regulation 1129/2011 Part C Group III (food colours with combined maximum limit).
After 1 August 2014 the aluminium lake is only permitted in food categories where a maximum limit on aluminium from lakes is explicitly stated in Part E of Regulation 1129/2011.

Typical Products
Baked goods and confectionery.

Number	Product
E160a(i)	Mixed carotenes
	CI orange 5
	Colour Index No: 75130
E160a(ii)	ß-carotene
	CI orange 5
	Colour Index No: 40800

Sources
Mixed carotenes are a mixture of natural products obtained by solvent extraction of edible plants and vegetables. ß-carotene is the majority constituent of this mixture.

The colour ß-carotene is a nature-identical pigment that is produced by chemical synthesis. Although both mixed carotenes and ß-carotene are oil-soluble colours, water-dispersible preparations are commercially available.

Function in Food

Mixed carotenes and ß-carotene are used to provide yellow and orange shades when used to colour foods.

The pure pigment has a high tintorial strength with an absorption maximum in the range 440–457 nm when measured in cyclohexane.

Benefits

ß-carotene is a widely distributed carotenoid with a long history of consumption by man. It is a precursor of vitamin A that thus enables the body to utilise it as pro-vitamin A. Its stability to heat and pH change is generally good. Typical dose applications calculated on the basis of the pure pigment are between 5 and 50 ppm.

Limitations

Carotenes are sensitive to oxidation especially when exposed to light. Foods and beverages coloured with carotenes frequently benefit from the protective addition of vitamin C and vitamin E derived antioxidants. It has an ADI of 0–5 mg/kg body weight allocated by JECFA and, in the EU is included in Part C Group II of Regulation 1129/2011, food colours authorised at *quantum satis*. In the USA it is authorised for use in food according to GMP.

Typical Products

Beverages, yellow fats, dairy products and flour confectionery.

Number	Product
E160b	Annatto
	Bixin
	Norbixin
	CI natural orange 4
	Colour Index No: 75120

Sources

The seeds of the tropical bush *Bixa orellana* L have long been used as a spice in Central and South America. The bush and the pigment extracted from them are both known as annatto. Bixin is the principle carotenoid pigment extracted from the seeds. Alkaline hydrolysis of bixin converts it from an oil-soluble colour to the water-soluble pigment norbixin.

Function in Food

Annatto is a naturally derived colour that may be used to provide orange shades in both lipid and aqueous food phases.

The pure pigment has a high tinctorial strength with characteristic absorption maxima of 470 nm/502 nm for bixin and 452 nm/482 nm in respect of norbixin.

Benefits
Stability of annatto to heat is excellent and it may generally be used in products throughout the acid pH range.
Typical dose applications calculated on the basis of the pure pigment are between 10 and 50 ppm.

Limitations
Annatto has an ADI of 0.065 mg/kg body weight allocated by JECFA and, in the EU, it is an approved colour according to Regulation 1129/2011 where it is permitted in a range of foodstuffs with individual maxima in each case. In the USA it is permitted in food according to GMP.
As a carotenoid, annatto is sensitive to oxidation especially when exposed to light.

Typical Products
Yellow fats, cheese, smoked fish, snacks and desserts.

Number   Product
E160c   Paprika extract
                                      Paprika oleoresin
                                      Capsanthin
                                      Capsorubin

Sources
Paprika colour is obtained from the sweet red pepper, *Capsicum annuum*, using a solvent extraction process to prepare an oleoresin. Paprika is well recognised as a spice and is a popular ingredient in many recipe dishes. Although the pigments are oil soluble, water-dispersible preparations are available commercially.

Function in Food
Paprika extract contains the oil-soluble carotenoid pigments capsorubin and capsanthin. It provides a deep orange hue and imparts a mild spice flavour when used in food products.
The pure pigment has a high tinctorial strength with an absorption maximum in the region of 462 nm when measured in acetone.

Benefits
Paprika colour derives from a spice that has a long history of consumption by man. Stability to heat and pH change is generally good and the mild spice note can be beneficial when it is used in savoury products.

Typical dose applications calculated on the basis of the pure pigment are between 10 and 50 ppm.

Limitations

Paprika pigments are carotenoids and are sensitive to oxidation especially when exposed to light. High dose levels may contribute an unacceptable flavour especially when used in mild flavoured sweet preparations. It does not have a specified ADI according to JECFA since it is considered that its use as a spice is self-limiting. It is an approved colour for use in foodstuffs in the EU in Pact C Group II of Regulation 1129/2011, food colours authorised at *quantum satis*. It is a permitted food colour in the USA according to GMP.

Typical Products

Soups, pickles, meat products, sauces, breadcrumbs and snack seasonings.

Number	Product
E160d	Lycopene
	CI natural yellow 27
	Colour Index No: 75125

Sources

Lycopene is a carotenoid obtained by solvent extraction of red tomatoes (*Lycopersicon esculentum* L.). It is an oil-soluble pigment, the commercial preparations of which consist of a mixture of carotenoids with lycopene being the principal constituent.

Function in Food

Lycopene provides orange and red colours to foods. The pure pigment has a high tinctorial strength with an absorption maximum in the region of 472 nm when measured in hexane.

Benefits

Lycopene is a natural colour derived by physical means from red tomatoes. Accordingly, it has a long history of consumption by man and normal dietary intake considerably exceeds that used for the purpose of colouration. It is unaffected by pH and exhibits good stability to heat. Typical dose applications calculated on the basis of the pure pigment are between 5 and 50 ppm.

Limitations

Lycopene is an oil-soluble carotenoid pigment. It is therefore sensitive to oxidation especially when exposed to light.
JECFA has allocated an ADI of "not specified". The colour is approved for use in the EU according to Regulation 1129/2011 in a wide range of foods subject to specific quantitative limits. In the USA, it is authorised for use in food according to GMP.

Typical Products
Soups, sauces, bakery wares, edible ices, processed cheese.

Number	Product
E 160e	β-apo-8′-carotenal (C30)
	CI food orange 6

Colour Index No: 40820

Sources
Although widely distributed in nature, commercial quantities of apocarotenal are chemically synthesised. The pure crystals, like β-carotene, are prone to oxidation, and the generally available form is as an oil suspension or water-dispersible powder.

Function in Food
β-apo-8′-carotenal imparts a yellow-orange to orange-red colour to foodstuffs. It has an absorption maximum of about 462 nm when measured in cyclohexane.

Benefits
Apocarotenal has good heat stability and the colour is unaffected by changes in pH. Its high colour intensity means that relatively low inclusion rates (5–15 ppm) of pure pigment are required.

Limitations
As a carotene, apocarotenal is prone to photo-oxidation, and products should be either protected from UV light or the formulation should include an anti-oxidant such as ascorbic acid. The carotenes, including apocarotenal, are not soluble in water but water-dispersible forms are available.
Like β-carotene, apocarotenal has vitamin A activity and has been allocated the same ADI of 5 mg/kg body weight by JEFCA. In the USA it is permitted for general use at a rate of 15 mg/lb, while in the EU it is included in Part C Group III of Regulation 1129/2011, food colours with combined maximum limit.

Typical Products
Soft drinks, confectionery, edible ices and desserts.

Number	Product
E161b	Lutein
	Mixed carotenoids
	Xanthophylls
	Tagetes

Sources
Xanthophylls are a class of carotenoids obtained by solvent extraction from edible fruits and plants including grass, lucerne (alfalfa) and marigolds (*Tagetes erecta*).

E 161b is an oil-soluble pigment consisting of a mixture of carotenoids with lutein being the principal constituent.

Function in Food
Oil-soluble and water-dispersible preparations are available commercially to provide a yellow colour to foods in either lipid or aqueous phases.
The pure pigment has a high tinctorial strength with an absorption maximum in the region of 445 nm when measured in chloroform/ethanol or hexane/ethanol/acetone solvent systems.

Benefits
Lutein is a natural colour derived by physical means from edible plants and fruits. It is unaffected by pH and exhibits good stability to heat. Typical dose applications calculated on the basis of the pure pigment are between 5 and 50 ppm.

Limitations
Xanthophyll pigments are carotenoids and are sensitive to oxidation especially when exposed to light.
Lutein has an ADI of "not specified" according to JECFA and, in the EU, is included in Part C Group III Regulation 1129/2011, food colours with combined maximum limit. EFSA has allocated an ADI of 1 mg/kg body weight

Typical Products
Cloudy citrus beverages, sugar confectionery, marzipan and mayonnaise.

Number	Product
E 161g	Canthaxanthin
	CI food orange 8
	Colour Index No: 40850

Sources
Canthaxanthin is found in nature in crustacea and as a result in salmon and trout flesh and flamingo feathers. Commercial canthaxanthin is chemically synthesised. The deep violet crystals are prone to oxidation and are insoluble in water. An oil-suspension and water-dispersible dry forms are available.

Function in Food
Canthaxanthin is used for colouring feedstuffs and some medicines.

Limitations
In the EU, canthaxanthin is included in Regulation 1129/2011 only because it is permitted in medicinal products in accordance with Directive 2009/35/EC and in feed. It is not permitted in foodstuffs. In the USA it is only permitted in feeds with maximum limits.

Number   Product
E 162   Beetroot red
                                          Beet red
                                          Betanin

Sources
Beetroot red is obtained from the roots of natural strains of red beet (*Beta vulgaris*) by pressing the crushed beet to express the juice or alternatively by aqueous extraction of shredded beetroots.
The main colouring principle consists of betacyanins of which betanin is the major component.

Function in Food
Beetroot red is used to impart a blue pink colour to foods. Commercial preparations are relatively low in respect of pigment content although this is partially balanced by the high tinctorial strength of the major pigment, betanin. The absorption maximum is in the region of 535 nm when measured in aqueous solution at pH 5.0.

Benefits
Beetroot red is a water-soluble colour obtained by physical means from a vegetable with a long history of consumption by man. Beet red provides a bright red hue that is relatively unaffected by pH. Typical dose applications calculated on the basis of the pure pigment are between 5 and 50 ppm.

Limitations
Beetroot red is fairly sensitive to heat, light and water activity and is extremely sensitive to sulfur dioxide. The effect of these limitations may be reduced if specially formulated products are utilised.
It has an ADI of "not specified" according to JECFA and is approved for use in foodstuffs in the EU in Regulation 1129/2011 Part C Group II, food colours authorised at *quantum satis*. In the USA it is permitted in foodstuffs according to GMP.

Typical Products
Ice cream, dairy products, dessert mixes and icings.

Number   Product
E 163   Anthocyanins
                                          Grape skin extract
                                          Grape colour extract
                                          Enocianina

Sources
Anthocyanins are the mainly red pigments that are responsible for the colours of many edible fruits and berries. They are usually obtained by aqueous

extraction, often using sulfurous acid. The major commercial source is grape skins but anthocyanins are also produced commercially from other edible materials including elderberries, red cabbage and black carrots.

Function in Food
Anthocyanins are naturally occurring pigments that are widely used to impart either red or purple shades to foods. The appearance of these water-soluble colours is dependent upon the pH of the product in which they are used. Colour hue progresses from red to blue as the pH increases and the anthocyanins are normally used in acidified products with a pH below 4.5.
The absorption maximum is in the region of 515/535 nm when measured in aqueous solution at pH3.0.

Benefits
Anthocyanins are water-soluble colours obtained by physical means from edible fruits and vegetables. Accordingly they have a long history of consumption by man and normal dietary intake considerably exceeds that used for the purpose of colouration. Typical dose applications calculated on the basis of the pure pigment are between 10 and 100 ppm.

Limitations
Because the stability, shade and colour intensity of anthocyanins are influenced by pH, they are not generally suited for colouring foods with a pH above 4.5. Some anthocyanins exhibit sensitivity to sulfur dioxide and protein but these limitations can usually be overcome by careful product selection.
They have an ADI according to JECFA of "not specified" and, within the EU, are included in Part C Group II in Regulation 1129/2011 as food colours authorised at *quantum satis.*

Typical Products
Soft drinks, jams, sugar confectionery, fruit toppings and sauces.

Number   Product
E 170   Calcium carbonate
                                                 CI pigment white 18

                                                 Colour Index No: 77220

Sources
Calcium carbonate is a mineral naturally occurring as chalk, limestone or marble. Food-grade calcium carbonate is taken from sources that are at least 99% pure calcium carbonate. Precipitated calcium carbonate is made by heating a mineral source to form calcium oxide, followed by reaction with water to form calcium hydroxide and then with carbon dioxide to form the calcium carbonate, which is purified by flotation.

Function in Food
Calcium carbonate is used as a colour, a source of carbon dioxide in raising agents, an anticaking agent, a source of calcium and a texturising agent in chewing gum.

Benefits
Calcium carbonate is readily available and inexpensive. It can be used in raising agents as it releases carbon dioxide both on addition of acid and on heating.

Limitations
Calcium carbonate is not a bright white colour and titanium dioxide is often preferred.
It has been allocated an ADI of "not specified" by EFSA and is authorised in Part C Group II of Regulation 1129/2011, food colours authorised at *quantum satis*. However, it is not permitted as a colour in foodstuffs in the USA.

Typical Products
Chewing gum and bread.

Number	Product
E 171	Titanium dioxide
	CI pigment white 6
	Colour Index No: 77891

Sources
Titanium dioxide is extracted from natural ores and milled to the correct particle size to provide optimum opacity and whiteness. It exists in three different crystalline forms, known as rutile, anatase and brookite. The anatase and rutile forms are permitted for use in foodstuffs. Processing conditions determine the form.

Function in Food
Titanium dioxide is a white powder that is used to colour foodstuffs, to provide opacity and to give a light background for other colourings.

Benefits
Titanium dioxide is the only true white colour permitted in the EU and the USA. It is insoluble in water and is stable to heat, light, acids and alkalis.

Limitations
Titanium dioxide is only available as a powder. It has been allocated an ADI of "not specified" by JEFCA (1969) and, in the EU, is included in Regulation 1129/2011 Part C Group II, food colours authorised at *quantum satis*. Titanium dioxide is permitted in the USA at less than 1.0% by weight of food.

Typical Products
Confectionery, edible ices and coffee whiteners.

Number	Product
E 172	Iron oxides and hydroxides
	Colour Index No. Iron oxide yellow: 77492
	Iron oxide red: 77491
	Iron oxide black: 77499

Sources
Iron oxides and hydroxides are produced by the controlled oxidation of iron in the presence of water. The colour range is from yellow through red to black, the precise colour being controlled by the details of the manufacturing process. Even within reds, there are shades from yellow red to blue red.

Function in Food
The iron oxides and hydroxides provide basic yellow, red or black colours to foodstuffs.

Benefits
The pigments are highly stable, particularly to light, and are unaffected by normal food ingredients. They are also suitable for products that are heat processed.

Limitations
The colours provided by iron oxides and hydroxides are dull and if used at high levels can interfere with metal-detection systems. The oxides and hydroxides are insoluble in water. They have been allocated an ADI of "not specified" in the EU (1975) and 0–5 mg/kg bodyweight by JECFA in 1979. In the EU, they are included in 1129/2011 Part C Group II, food colours authorised at *quantum satis*.

Typical Products
Fish paste, canned goods, confectionery and pet food.

Number	Product
E 173	Aluminium
	Colour Index No: 77000

Sources
Food grade aluminium is at least 99% pure aluminium.

Function in Food
Aluminium powder is used to colour small decorative pieces used on cakes to give a bright metallic shine.

Limitations
Within the EU the use of aluminium in food is limited under Regulation 1129/2011 Annex II to external coating of sugar confectionery for the decoration of cakes and pastries. However, under Regulation 380/3012, this use is to be discontinued in February 2014.

Number	Product
E 174	Silver

Colour Index No: 77820

Sources
Food grade silver is 99.5% pure silver. Silver is only available as bars or wire; there is no silver product analogous to gold leaf.

Function in Food
Silver is used to cover chocolates and to provide coloured particles in liqueurs. Very little silver is used in this way.

Limitations
Silver tarnishes readily in air, powdered silver particularly so. Within the EU the use of silver in food is limited under Regulation 1129/2011 to external coating of confectionery, chocolates and liqueurs.

Number	Product
E 175	Gold

Colour Index No:77480

Sources
Food grade gold is 99.99% gold, which is permitted to be mixed with no more than 7% silver or 4% copper. The addition of these metals improves the malleability of the gold so that it can be hammered into gold leaf. Gold mixed with copper has an orange tinge and that mixed with silver a greenish tinge.

Function in Food
Gold is used as gold leaf to wrap chocolate confections, and as tiny pieces to colour confectionery and liqueurs.

Limitations
Within the EU the use of gold in food is limited under Regulation 1129/2011 Annex II to external coating of confectionery, chocolates and liqueurs.

Number   Product
E 180   Litholrubine BK
                                          FD&C red no. 6

Sources
Litholrubine is a synthetic azo dye that is water insoluble. Its absorption
maximum in dimethylformamide is 442 nm.

Function in Food
Litholrubine is a bright red colour used to colour the edible rind of cheeses. Its
major use is in cosmetics.

Limitations
Within the EU litholrubine is permitted under Regulation 1129/2011 Annex II
only in edible cheese rind.

Number   Product
E 200   Sorbic acid
E 202   Potassium sorbate
E 203   Calcium sorbate

Sources
Sorbic acid is the *trans, trans*, isomer of 2,4,-hexadienoic acid. It occurs
naturally in the unripe fruits of the mountain ash, *Sorbus aucuparia L.* and in
some wines. The material of commerce is synthetic.

Function in Food
Sorbic acid and sorbates are used mainly as food preservatives. They are
effective against a wide range of micro-organisms, particularly yeasts and
moulds, including organisms responsible for mycotoxin formation. Among
bacteria, aerobic bacteria are affected the most. Sorbic acid and sorbates are
often used in synergistic combinations with other preservatives.

Benefits
Sorbic acid and its derivatives can be used in many different types of product
with a wide range of pH values. They do not interact with other food
ingredients, and are neutral in both taste and flavour. Dry sorbates are very
stable.

Limitations
Within the EU, sorbic acid and the sorbates are permitted in Regulation
1129/2011, in a range of products with individual maxima in each case.
Typical usage levels are 2000 ppm in solid food and 300 ppm in beverages.
Sorbates must not be used in products in whose manufacturing fermentation
plays an important role because they inhibit the action of yeast. If potassium

sorbate is combined with other preservatives, care must be taken that no calcium ions are present, as this brings about precipitation. Therefore, for combinations with potassium sorbate, sodium propionate should be used instead of calcium propionate in order to obtain good synergistic action.

Sorbate is believed not to have an effect against oxidative and enzymic browning and should be combined with sulfur dioxide or heat pasteurisation to inhibit browning.

Typical Products
Baked goods, nonalcoholic beverages, cheese, dairy products, meat products and in fungistatic packing material.

Number	Product
E 210	Benzoic acid
E 211	Sodium benzoate
E 212	Potassium benzoate
E 213	Calcium benzoate

Sources
Benzoic acid is produced by the oxidation of toluene.

The salts of benzoic acid are made by reacting the acid with the appropriate hydroxide. Sodium benzoate is the most common of the three salts used in commerce although its use has decreased in recent years.

Benzoic acid may be naturally present in some fermented products as a result of the fermentation process.

Function in Food
The benzoates are used as preservatives to inhibit the growth of yeasts and moulds and have been used since the early 1900s. They have less effect against bacteria. They are often synergistic with other preservatives, such as sorbates, and are used in conjunction with sulfur dioxide, which itself inhibits enzyme action and browning.

Benefits
The benzoates are readily soluble in water and readily available. Sodium benzoate is the most common product; potassium benzoate being used where a lower sodium content is required. The benzoates are used in acid products, where they are present as benzoic acid. The acid itself is insoluble in water but moderately soluble in oils.

Limitations
The benzoates have a distinctive flavour, which limits the concentration at which they can be used. They cannot be used in yeast-raised flour products because they inactivate the yeast. Benzoic acid is only slightly soluble in water.

In liquid products that include ascorbic acid, under certain conditions, an interaction can occur that produces benzene in levels of a few parts per billion. Benzoic acid and benzoates are permitted in the EU in Regulation 1129/2011, either alone or in combination with other preservatives in a range of products with individual limits in each case and in Regulation 570/2012 in alcohol-free wines to a maximum of 200 mg/l. They have been allocated a total ADI for all benzoates of 0–5 mg/kg body weight, expressed as benzoic acid.

Typical Products
Soft drinks, alcohol-free wines.

Number	Product
E 214	Ethyl p-hydroxy benzoate
E 215	Ethyl p-hydroxy benzoate sodium salt
E 218	Methyl p-hydroxy benzoate
E 219	Methyl p-hydroxy benzoate sodium salt

Sources
The esters of p-hydroxy benzoic acid are produced by reacting the respective alcohols with p-hydroxy benzoic acid. The acid itself is made by reaction of potassium phenate with carbon dioxide under pressure at high temperature.

Function in Food
The p-hydroxy benzoate esters are preservatives against yeasts and moulds, but are less effective against bacteria, especially Gram-negative species. Their effectiveness is dependent on the individual species and they are often used in combination with sorbic or benzoic acid. They are rarely used in foods and their main uses are in cosmetic and personal care products.

Benefits
The esters tend to be more effective antimicrobial agents than benzoic and sorbic acids and their effect increases with chain length of the ester group. Unlike some other preservatives they are effective in water at pH levels from neutral to mildly acid. They are moderately soluble in oils. They tend to be used in combination as their effect is additive but their taste is not.

Limitations
The taste of the esters is detectable even at low levels in food products so the rate of use is self-limiting. They are only slightly soluble in water. They are permitted in the EU in Regulation 1129/2011 in a very limited range of products with individual maxima.

Typical Products
None known.

Number	Product
E 220	Sulfur dioxide
E 221	Sodium sulfite
E 222	Sodium bisulfite, sodium hydrogen sulfite
E 223	Sodium metabisulfite
E 224	Potassium metabisulfite
E 226	Calcium sulfite
E 227	Calcium bisulfite, calcium hydrogen sulfite
E 228	Potassium bisulfite

Structure

All the substances that are listed as E 220–E 228 are equivalent when they are present in food. E 221–E 228 are all salts of sulfurous acid that is formed when sulfur dioxide (E 220) is dissolved in water. The actual species that are present in food depend upon the nature of the food and not upon the chemical form of the additive. It is only in the most acid of foods, *e.g.* lemon juice and wines, that significant levels of E 220 itself occur. Otherwise, the preservative is converted, upon addition to food, into ionic species, mostly hydrogen sulfite and sulfite ion, and into ionic reaction products, all of which are nonvolatile. The reason for the relatively large number of "equivalent" substances is technological. Thus, E 220 would be used as the additive of choice when fruit is fumigated or when it is desired to use the substance as an acidulant as well as a preservative. E 223 and E 224 are particularly stable when stored or handled in the factory environment. On the other hand, E 226 is relatively insoluble in water and would be used in situations in which solubility must be minimised. The most widely used form of this preservative is sodium metabisulfite, E 223. The term "sulfur dioxide" is used conventionally in the food industry to refer collectively to all these additives, because the recognised methods of analysis convert the additive, in whatever form it is, into sulfur dioxide gas. Legal specifications refer to the mass of sulfur dioxide released upon the analysis of 1 kg of food. However, the individual substances need to be listed with E-numbers because the same mass of each is equivalent to a different amount of sulfur dioxide, and each substance has defined purity criteria. In this section, the term sulfites will be used to refer collectively to substances E 220–E 228, to avoid confusion with the specific substance, sulfur dioxide.

Sources

All substances in the range E 221–E 228 are obtained by the addition of sulfur dioxide to the appropriate alkali (sodium, potassium or calcium hydroxide) until the stoichiometric amount has undergone reaction, and the product is then crystallised. Sulfur dioxide is produced synthetically by burning sulfur or, for example, various metal sulfides. It is the starting material for the production of sulfuric acid and so is available cheaply and in a pure state.

Function in Food

Sulfites are the most versatile of all food additives. They have been used in foods since the times of the ancient Romans and Greeks, and are important ingredients in certain traditional foods. Their listing as food preservatives indicates that a primary function is to act as an antimicrobial agent. In this role, sulfites are most effective in acid foods in which the efficacious agent is sulfur dioxide itself. However, sulfites are also added to food to control chemical spoilage, in which capacity they play a unique role. The most well-known applications are the control of enzymic browning at the cut or damaged surfaces of plant foods, and nonenzymic browning of sugars or vitamin C when foods are processed thermally or stored. Sulfites inhibit most forms of enzymic spoilage in foods, *e.g.* those involving oxidising enzymes such as peroxidases and lipoxygenases, which can otherwise cause off-flavours. They prevent oxidative rancidity when unsaturated fats are oxidised nonenzymically in plant foods, and help to preserve vitamins A and C. E 223 is used exceptionally as a processing aid to modify the physical characteristics of wheat flour for biscuit manufacture. Sulfites are used to bleach cherries before they are coloured artificially. Contrary to some belief, sulfites do not restore the colour of discoloured meat, but help retain the red colour when used in sausages. The effect of the sulfites on any food product is thus seen to be complex and it is recognised that the sensory properties of foods treated with this range of additives differ uniquely from those of the untreated foods. This includes a contribution to the characteristic taste of some foods from sulfur dioxide.

Benefits

As an antimicrobial agent, sulfites are used widely to preserve fermented and nonfermented beverages. Their primary purpose here is to prevent spoilage in storage and after the beverage container has been opened. In wine-making, the resistance of specific yeasts to sulfur dioxide is used to select against "wild" yeasts in fermentation and subsequent storage. It is also thought to contribute to the characteristic dry taste of some white wines. The additive is used against salmonellae and the spoilage yeasts in meat products such as sausage, thus extending the shelflife of this food. As an antibrowning agent it is used in food production to control enzymic browning after fruit and vegetables are peeled before processing. For this reason, some catering packs of "fresh" peeled potatoes are treated with this additive. It is essential in the production of pale-coloured dried fruit such as apricots, peaches and sultanas. Vegetable dehydration depends critically on sulfites to prevent discolouration during production. Subsequently, sulfites allow dehydrated fruits and vegetables to be stored for long periods of time without specialised storage requirements. In these respects, there are no practical alternatives. As an enzyme inhibitor, sulfites prevent the formation of off-flavours, particularly those that arise from the action of oxidising enzymes on fats. As antioxidants, they help to extend the shelflife of dehydrated vegetables such as potato and increase the retention of vitamins A and C. They also increase the stability of natural food colours such as the carotenoids (*e.g.* in dehydrated carrot, peppers, tomato). Sulfites are unique in their control of the staling of beer. As a processing aid, they allow accurate control of the physical properties of wheat flour for biscuit manufacture to ensure a consistent product. A combination of these functions allows fruit to be stored in pulp for many months for jam manufacture without the need for freezing.

Limitations

Within the EU sulfur dioxide and the sulfites are permitted in Regulation 1129/2011. Sulfites are a normal part of the human metabolism even when there is none of the additive in the diet. Whilst the human body is remarkably well able to metabolise and detoxify this additive when ingested, that which is inhaled (as sulfur dioxide gas) can cause an adverse reaction (sometimes severe) in a small number of individuals, particularly those who suffer from asthma. Small concentrations of this gas are present in the headspace above foods in which the additive is present, the highest concentrations being found in acidic food products. There is some concern that individuals who consume large amounts of wine, or who have a diet biased towards foods treated with sulfites, can exceed the acceptable daily intake of ingested additive, but there is no known adverse effect arising from such excessive consumption. The classical antinutritional behaviour of the additive is that it destroys vitamin B_1 in food, but this is not thought to give rise to vitamin deficiency. A technical limitation is that the amount of the additive present in most foods decreases with time as a result of the many chemical reactions that are required for it to exert its preservative effect. This means that food treated in this way has a limited shelflife. On the other hand, there are toxicological implications arising from the reaction products. Evidence suggests that a major reaction product formed from sulfite is metabolically inert and, therefore, harmless. Sulfites may also be present at a very low level in foods where the additive does not serve a technological function, as a result of carry-over in the ingredients or from preprocessing operations.

Typical Products

Soft drinks and fruit juices; fermented drinks, including beer, wine, cider and perry; dehydrated vegetables; dehydrated fruits; peeled potatoes; maraschino cherries; sausages and burgers; jam (as a result of use in fruit pulp); and biscuits.

Number	Product
E234	Nisin

Sources

Nisin is an antimicrobial peptide (small protein) or "bacteriocin" produced by certain strains of the lactic acid bacterium *Lactococcus lactis* subsp. *Lactis*. Commercial preparations, standardised to 2.5% nisin (one million international units per gram), are prepared by the controlled fermentation of nisin-producing *L. Lactis* subsp. *lactis* strains in a milk-based medium, recovery and drying of the nisin and blending with salt. Dry nisin preparations are very stable, providing storage is below 25 °C.

Function in Food

Nisin is used as a food preservative and shows strong antimicrobial activity against Gram-positive bacteria but no activity against Gram-negative bacteria, yeasts and moulds. Amongst Gram-positive bacteria, nisin is particularly active against the spore-forming genera *Clostridium* and *Bacillus*. Both spores and vegetative cells are sensitive to nisin, although spores are usually more sensitive than their vegetative cell equivalent. Other nonspore-forming bacteria that are sensitive to nisin are lactic acid bacteria, and *Listeria monocytogenes*.

Benefits

Nisin is an important preservative in foods that are pasteurised but not fully sterilised, since pasteurisation kills Gram-negative bacteria, yeasts and moulds but not bacterial spores. In heat-processed foods, nisin can be used to allow a reduction in the heat-processing regimes, which has the benefit of protecting food against heat damage, thus improving nutritional content, flavour, texture and appearance, and providing an energy saving.

As an antimicrobial agent against lactic acid bacteria, nisin has applications in low-pH (high acid) foods such as sauces and salad dressings, and for the control of spoilage lactic acid bacteria in beer, wine and spirit manufacture. In certain foods, such as ricotta, feta and cottage cheese, it can be used to inhibit *Listeria monocytogenes*.

After consumption in food, nisin is degraded by the digestive protease enzymes; thus no passage or accumulation of nisin will occur in the body. There are no reported allergic responses of human beings against nisin in food.

Limitations

Nisin is recognised as being of very low or no toxicity and has GRAS status for the intended uses in the USA.

In EU Regulation 1129/2011, nisin is permitted in a limited range of products – clotted cream, certain cheeses and cheese products, pasteurised liquid egg, semolina and tapioca puddings, with individual maxima. Nisin has been allocated an ADI of 0–0.13 mg/kg body weight by the SCF. Usage levels range from 0.5 to 15 mg/kg of foodstuffs.

Typical Products

Processed cheese products.

Number	Product
E 235	Natamycin
	Pimaricin
	Tennectin

Sources

Natamycin is a natural antimicrobial produced by *Streptomycetes* bacteria found in soil worldwide. Producing organisms are typified by *Streptomycetes natalensis*, from Natal, South Africa, where it was isolated in 1955. Commercial preparations are made by the controlled fermentation of dextrose-based media by selected *Streptomycetes* strains. Dried natamycin recovered from the fermentation broth is white to cream-coloured, has little or no odour or taste, and in the crystalline form is very stable. Solubility in water and most organic solvents is low.

Function in Food
Natamycin is used as a food preservative. It shows strong activity against yeasts and moulds, but no activity against bacteria or viruses. Its effect is predominantly fungicidal.

Benefits
Natmycin is usually used as a surface treatment to prevent the growth of yeasts and moulds where its low solubility ensures that it remains on the surface and therefore active at the site where most yeasts and moulds will occur. Natamycin does not interact with other food ingredients, and imparts no off-flavours to food.

Limitations
Natamycin has been allocated an ADI of 0–0.3 mg/kg and given GRAS status for the intended uses in the USA.
Included in EU Regulation 1129/2011, Natamycin is only permitted in a limited range of products – surface treatment of hard cheese, cheese products and dried cured sausages.

Typical Products
Cheese coatings, sausages.

Number	Product
E 239	Hexamethylene tetramine

Sources
Hexamethylene tetramine is made by reacting formaldehyde with ammonia and purifying the product.

Function in Food
Hexamethylene tetramine is a preservative that works by releasing formaldehyde in acid conditions. The formaldehyde prevents "late blowing" in hard cheese by inhibiting the growth of the bacteria that cause this defect.

Limitations
Hexamethylene tetramine has a slightly sweet taste with a bitter aftertaste.

Under Regulation 1129/2011 it is permitted only in Provolone, an Italian hard cheese, to a maximum residual amount of 25 mg/kg, measured as formaldehyde.

Number	Product
E 242	Dimethyl dicarbonate (DMDC)

Sources
Dimethyl dicarbonate (DMDC) is manufactured by chemical synthesis with specially designed extraction and distillation steps to obtain the required purity.

Function in Food
DMDC is used as an antimicrobial agent for the cold sterilisation of beverages.

Benefits
Even at very low concentrations, DMDC is very effective against typical beverage-spoiling micro-organisms, such as fermentative yeasts, mycoderma and fermentative bacteria. At higher concentrations, it destroys a large number of bacteria, wild yeasts and moulds.
Shortly after DMDC has been added to the beverage, it breaks down completely by hydrolysis into minute amounts of microbiologically inactive products.
Sensory tests, and many years of experience have shown that DMDC does not influence the taste, colour or odour of beverages.

Limitations
DMDC is permitted in the EU, in Regulation 1129/2011, only in a range of beverage products.
As the beverage temperature has a major impact on the hydrolysis rate, the beverage should be cooled down to a maximum of 20 °C. This will slow down the decomposition of DMDC, thus prolonging its antimicrobial action. Low temperatures, therefore, support the efficacy of DMDC and its economical use.

Typical Products
Nonalcoholic drinks, ready-to-drink tea, juice concentrates.

Number	Product
E 249	Potassium nitrite
E 250	Sodium nitrite

Sources
Sodium nitrite is made when a mixture of the oxides of nitrogen is passed into sodium hydroxide. The nitrite crystallises on cooling a concentrated solution. It is available as both solution and crystals. The major uses of sodium nitrite are in the chemical industry.

There is very little production of potassium nitrite for the food industry. By far the majority of nitrite in commerce is the sodium salt.

Sodium nitrite is naturally present in some plant foods including lettuce and spinach.

Function in Food
The nitrites are used as preservatives, preventing the growth of pathogenic organisms in meat. They have no action against yeasts or moulds.

Benefits
The nitrites are one of the few substances available for preserving cured meats by inhibiting the growth of anaerobic bacteria such as *Clostridium botulinum*. The antimicrobial effect is enhanced when the nitrite is added before the food is heat processed. Nitrites have the added benefits of preserving the red colour of meat (by reacting with the myoglobin) and assisting in the development of the typical "cured" flavour. They also act as antioxidants and prevent the formation of "warmed-over" flavour, which develops when cooked meat is kept exposed to air.

Limitations
In the EU, nitrites are included in Regulation 1129/2011 where they are permitted as preservatives only in specified processed meat products.

The nitrites have been allocated an ADI by JECFA of 0–0.06 mg/kg body weight (expressed as sodium nitrite).

Typical Products
Bacon, ham and traditional cured meat products.

Number	Product
E 251	Sodium nitrate
E 252	Potassium nitrate

Sources
Sodium nitrate is found in nature as Chile saltpetre. It is also produced as a byproduct of the production of sodium nitrite.

Potassium nitrate is also found as a natural product, but is produced commercially by reacting potassium carbonate with nitric acid.

The major use of potassium nitrate is in agriculture.

Nitrates are naturally present in a number of plant foods including spinach and lettuce.

The majority of nitrate in the diet arises from these foods and, in the EU, the maximum level of nitrate permitted in them is set by Regulation 1881/2006.

Nitrate is also present in the domestic water supply in many countries and in the EU this is controlled by measures laid out in Directive 98/83 EC.

Function in Food

The nitrates, particularly the sodium salt, have been used for at least two thousand years as a preservative, often in combination with the nitrite and salt. They work by being converted into nitrites (see E 249/250) in the food by enzymes that are present in the food and in bacteria.

Benefits

Use of the nitrates is one of the few methods of inhibiting the growth of anaerobic bacteria, such as *Clostridium botulinum*. This makes them particularly useful in the production of cheese, cured meats and pickled fish.

Limitations

Nitrates are permitted in the EU within Regulation 1129/2011 in a limited range of products – a range of cheeses, cheese products and cheese analogues, nonheat-treated processed meat and traditional cured meat products. The nitrates have been allocated an ADI by both JECFA and the SCF of 0–5 mg/kg body weight expressed as sodium nitrate (or 0–3.7 mg/kg body weight expressed as nitrate ion).

Typical Products

Cured ham, salami and cheese.

Number	Product
E260	Acetic acid

Sources

Vinegar is essentially a solution of 5 to 10% acetic acid in water. The original source of vinegar was the accidental bacterial oxidation of wine but this has been turned to advantage and today many varieties of vinegar are available derived from different sources of alcohol and flavoured with various herbs.

More concentrated solutions of acetic acid are manufactured industrially by oxidation of ethanol or hydrocarbons.

Function in Food

Acetic acid is naturally present in many foods and has been used for thousands of years as a preservative in pickles. In ancient Rome it was used in mixtures with salt, wine or honey. It is still used as a preservative, its effect resulting from the decrease in pH. The addition of salt increases its effectiveness, mainly by lowering the water activity.

Benefits

Acetic acid is used as a preservative largely in traditional products where its flavour contributes to the overall flavour in, for example, pickles, sauces and salad dressings.

Acetic acid is more effective against food-spoilage organisms than would be predicted from the pH and it is effective at higher pH than are other acids. It is synergistic with lactic and sorbic acids.

Limitations
Acetic acid is more effective against yeasts and moulds than against bacteria; lactobacilli being particularly resistant. In many products it is used in conjunction with other acids, preservatives or preservation methods, such as pasteurisation, to provide additional protection. Its readily recognised flavour limits the applications to savoury products.

In the EU, acetic acid is included in Group 1 Part C of Regulation 1129/2011, additives permitted at *quantum satis*. It has GRAS status in the USA.

Typical Products
Pickling liquids, marinades, sauces, salad dressings and mayonnaise.

Number	Product
E261	Potassium acetate

Sources
Potassium acetate is made by the reaction of acetic acid and potassium carbonate.

Function in Food
Potassium acetate is used as an acidity regulator and buffer.

Benefits
Potassium acetate is used to modify the flavour of products acidified with acetic acid. Its only advantage over sodium acetate is in products in which sodium content needs to be reduced.

Limitations
Potassium acetate has GRAS status in the USA and, in the EU, is included in Group 1 Part C of Regulation 1129/2011, additives permitted at *quantum satis*.

Typical Products
None known.

Number	Product
E 262	Sodium acetates
	(i) sodium acetate
	(ii) sodium diacetate

Sources
Sodium acetate is produced by reaction of acetic acid with sodium hydroxide. When acetic acid and sodium acetate are mixed in equimolar proportions and allowed to crystallise, the sodium acetate crystallises with acetic acid of crystallisation. This material is called sodium diacetate.

Function in Food
Sodium acetate is used as an acidity regulator and buffer.
Sodium diacetate provides a solid source of acetic acid for dry goods.

Benefits
Sodium acetate acts as a buffer and modifies the taste of acetic acid, softening the sharpness of the acid and making it more palatable. It is readily soluble in water.
Sodium diacetate has some specific uses as a source of acetic acid, for example in bread production, where it is used to protect against ropiness and against some moulds. It is also used as a flavouring in dry products, particularly to impart the flavour of vinegar.

Limitations
The sodium acetates have GRAS status in the USA and in the EU are permitted in Group 1 Part C of Regulation 1129/2011, additives permitted at *quantum satis*. Sodium diacetate should be stored in well-sealed containers as it loses acetic acid on storage.

Typical Products
Bread, flavoured snacks and instant soups.

Number   Product
E263   Calcium acetate

Sources
Calcium acetate is made by the reaction of acetic acid and calcium hydroxide.

Function in Food
Calcium acetate is used as an acidity regulator and as a source of calcium ions.

Benefits
Calcium acetate is readily soluble in water. It can be used to modify the flavour of products acidified with acetic acid or to provide a soluble source of calcium either for fortification or for reaction with alginates.
Calcium acetate is also used to protect against ropiness and against moulds in flour products but is less effective than sodium diacetate.

Limitations
In the EU, calcium acetate is included in Group 1 Part C of Regulation 1129/2011, additives permitted at *quantum satis*. It has GRAS status in the USA. It is more effective against moulds at lower pH levels.

Typical Products
Bread, and gelling mixtures such as vegetable gelatin.

Number	Product
E 270	Lactic acid

Sources

Lactic acid can exist in both a D and an L form. In the human body it is predominantly present as the L form. It is commercially available as both the L form, (made by fermentation of sugar or glucose) and as a racemic mixture of L and D forms (made synthetically). No dairy-based lactic acid is commercially available.

Function in Food

The main functions of lactic acid in the food industry are as an acidity regulator and a preservative.

Benefits

Lactic acid has a mild acidic flavour and tastes less acid at a given pH than other acids. As a consequence it is used where pH reduction is required to inhibit bacterial growth but a strong taste is not appropriate, for example as a preservative in pickled vegetables, low-fat mayonnaise and cooking sauces. It is effective against both spoilage and foodborne pathogenic bacteria. A combination of lactic acid and acetic acid (E 260) is used to inhibit the growth of yeasts.

Lactic acid is used as a pH regulator in dairy products and beverages including yogurts, cheese and in fruit flavour products where other organic acids would be too overpowering.

It is used to inhibit browning in confectionery products, to prevent the degradation of sugar and gelling agents and to mask off-flavours from intense sweeteners.

A particular use is in the decontamination of meat carcasses.

Limitations

Lactic acid is included in Group 1 Part C of Regulation 1129/2011, additives permitted at *quantum satis*, and has GRAS status in the USA.

Typical Products

Processed cheeses, pickled vegetables, salads, low-fat mayonnaise, soft candies, sour dough bread.

Number	Product
E 280	Propionic acid (propanoic acid)

Sources

Propionic acid is a naturally occurring short-chain fatty acid similar to acetic acid. It is liquid at room temperature and is a normal constituent of human body fluids. It can be produced by *Propionibacterium* from lactic acid and also by various methods involving the oxidation of propionaldehyde (a byproduct of fuel synthesis and wood distillation).

Propionic acid occurs naturally in ripe Swiss and Jarlsberg cheese at levels as high as 1%; it is also present in the rumen of ruminant mammals. It is digested and metabolised as a fatty acid in humans.

Function in Food

Propionic acid has a preservative effect as a mould inhibitor, and is active against many mould species, including *Aspergillus*, *Penicillium*, *Mucor* and *Rhizopus* – common spoilage organisms in bakery goods. It has a limited inhibitory effect on many yeast species, although it inhibits bakers' yeast species. Its inhibitory effect on bacteria is limited to retarding the growth of *Bacillus subtilis* (rope) in bread.

Benefits

The use of propionic acid extends the mould-free shelflife of bakery products and cheese and cheese products. It can also prevent blowing of canned frankfurters without affecting their flavour. Propionic acid is more effective at higher pH values, up to pH6, allowing preservation at higher pH levels than would otherwise be possible. This results in improved sensory quality of food.

Limitations

The use of propionic acid and propionates is controlled in the EU by Regulation 1129/2011 where they are permitted in a limited range of products with individual limits in each case. Propionic acid and the propionates have been allocated a joint ADI of 0–6.0 mg/kg bodyweight expressed as propionic acid.

High levels of propionic acid may create bitter, cheesy flavours and can reduce the activity of bakers' yeast. When using propionic acid in yeast-leavened products, the yeast level in the formulation should be increased and proof times may need to be extended.

Typical Products

Yeast and chemically leavened bakery products, prepacked and part-baked bread, cheese and cheese products.

Number	Product
E 281	Sodium propionate
E 282	Calcium propionate
E 283	Potassium propionate

Sources

The propionates are white, free-flowing, water-soluble salts manufactured by reacting propionic acid with carbonates or hydroxides. They are readily digested and metabolised in the body.

Function in Food

The propionates yield the free acid in the product to provide preservative action. They are active against many mould species, including *Aspergillus*, *Penicillium*, *Mucor* and *Rhizopus* – common spoilage organisms in bakery goods. They have a limited inhibitory effect on many yeast species, although they inhibit bakers' yeast species. The inhibitory effect on bacteria is limited to retarding the growth of *Bacillus subtilis* (rope) in bread.

Benefits

The propionates are used to extend the mould-free shelflife of bakery products, cheese and cheese products. They blend well with other ingredients and do not alter the colour taste or texture at normal usage levels. Being powders, they are easier to use than the liquid propionic acid.

The sodium and potassium salts are recommended for use in chemically leavened products because the calcium propionate may interfere with some chemical leavening agents.

Limitations

When using propionic acid or propionates in yeast-leavened products, the yeast level in the formulation should be increased and proof times may need to be extended.

Propionic acid and the propionates are controlled in the EU by Regulation 1129/2011 where they are permitted in a limited range of products. They have been allocated a joint ADI of 0–6.0 mg/kg body weight expressed as propionic acid.

High levels of propionates may create bitter, cheesy flavours and can reduce the activity of bakers' yeast.

Typical Products

Yeast and chemically leavened bakery products, prepacked and part-baked bread, cheese and cheese products.

Number	Product
E 284	Boric acid
E 285	Sodium tetraborate (borax)

Sources

Borax is a natural mineral that is mined and purified. Boric acid is produced by reaction of borax with sulfuric acid followed by purification and crystallisation.

Function in Food

Borax was first recommended as a preservative in 1775. It has been used for many years as a household disinfectant. Both the acid and sodium salt are effective against yeasts and, to a much lesser extent, against moulds and bacteria.

Benefits
Boric acid has a very low dissociation constant so it is largely undissociated, and thus effective, even at neutral pH, where carboxylic-acid-based preservatives have less effect. It is also water soluble and for many years was used in margarine and butter as it stayed in the aqueous phase.

Limitations
In the EU, under Regulation 1129/2011, boric acid and sodium tetraborate are currently only permitted in caviar to a maximum of 4 g per kilogram.

Number	Product
E 290	Carbon dioxide

Sources
Carbon dioxide makes up approximately 0.1% of the air that we breathe. It is a colourless, odourless, nonflammable gas, soluble in water to give an acidic solution. It is prepared from flue gases or as a byproduct from production of ammonia or hydrogen. It is then purified and liquefied. It is also available as a solid known as dry ice.

Function in Food
Carbon dioxide is used for the carbonation of beverages in the brewing and soft drinks industries and as a modified-atmosphere packaging gas.
In the latter role it has a powerful inhibitory effect on the growth of bacteria, being particularly effective against Gram-negative spoilage bacteria, such as *Pseudomonas* species. Carbon dioxide acts by forming a mild carbonic acid on the surface of the product, lowering the pH and producing an environment relatively hostile to bacteria It can also act as a powerful inhibitor of mould growth.

Benefits
The use of carbon dioxide in packaging extends the shelflife of products in terms of both food safety and quality. It is used on its own or in combination with other packaging gases, depending on product and pack format.

Limitations
Carbon dioxide is very easily absorbed into fats and is very soluble in water, particularly as the temperature decreases, so in a retail pack of a food product with a high water content there may be sufficient absorption to create a partial vacuum in the pack, causing it to distort or collapse. Under these circumstances, it may be advisable to incorporate a less-soluble gas (*e.g.* nitrogen) into the pack atmosphere along with the carbon dioxide to avoid this collapse, although this may have the effect of reducing the shelflife of the product. Indeed, combinations of carbon dioxide, nitrogen and oxygen are used to optimise the shelflife of a number of products in terms of appearance and safety.

Carbon dioxide has a relatively high transmission rate through packaging films compared with some other gases, so for the best extended shelflife the packaging must have good barrier properties.

As it is denser than air and toxic, care must be taken in its use in confined or low-lying working environments.

Carbon dioxide has GRAS status in the USA and, in the EU, is included in Group1 Part C of Regulation 1129/2011, additives permitted at *quantum satis*.

Typical Products

Carbonated beverages, part baked products (baguettes, rolls), hard cheese, fresh salads.

Number	Product
E296	Malic acid

Sources

Malic acid occurs naturally in many fruits including apples, peaches and cherries. The material of commerce is manufactured industrially from maleic anhydride.

Function in Food

Malic acid is used to provide acidity and to a much lesser extent to chelate metal ions, from hard water or in wine.

Benefits

Malic acid tastes less sharply acid than citric acid and is used to provide an acid taste that is less immediate but persists longer. It is used alone or in combination with other acids to give a range of acid impacts. It is particularly useful in product formulations that use intense sweeteners. It has a lower melting point than citric acid, which is beneficial in the manufacture of boiled sweets.

Limitations

In the EU, malic acid is included in Group 1 Part C of Regulation 1129/2011, additives permitted at *quantum satis*. It is recognised as GRAS in the USA.

Typical Products

Fruit drinks, boiled sweets, chewing gum, sorbets, jams and sweet and sour sauces.

Number	Product
E297	Fumaric acid

Sources

Fumaric acid occurs in many plants but is manufactured by fermentation or by isomerisation of maleic acid.

Function in Food
Fumaric acid provides an acid taste to products.

Benefits
The acid taste of fumaric acid complements and smoothes the acidity of other acids.
Fumaric acid is used in powdered products because it is only slightly hygroscopic.

Limitations
Fumaric acid is permitted in the EU in Regulation 1129/2011 in a number of classes of products, including fillings and toppings for bakery products, sugar confectionery, desserts and powdered dessert mixes, chewing gum and instant powders for fruit, tea or herbal tea based drinks, all with specified limits. It is also permitted in some wines. Its use is limited in the US to jellies and fruit preserves.
It is poorly soluble in water and dissolves slowly.
Fumaric acid has been allocated an ADI of 0–6 mg/kg body weight.

Typical Products
None known.

Number	Product
E 300	Ascorbic acid (vitamin C)
E 301	Sodium ascorbate
E 302	Calcium ascorbate

Sources
Ascorbic acid occurs naturally in most fruits and vegetables, notably in members of the citrus family. It is present in rose hips, from which it is extracted, but most of the material in commerce is made industrially by a six-step process starting with glucose. Sodium ascorbate is the sodium salt of ascorbic acid.

Function in Food
In solution, ascorbate is easily oxidised to dehydroascorbate. Ascorbic acid is added to foodstuffs that will be exposed to oxygen during storage so that it will be oxidised in place of the other ingredients of the food. This extends shelflife and preserves flavour. The oxidation reaction is readily reversible and some formulations benefit from this.

Benefits
Because ascorbic acid readily reacts with oxygen it is very effective at protecting other ingredients from oxidation. It is synergistic with other antioxidants and has little flavour. The antioxidant property is used to reduce discoloration in

canned fruit and vegetables and in fruit purées caused by polyphenol oxidase. It is also used in meat products to enhance colour formation and reduce the formation of nitrosamines. Ascorbates are used to increase the shelflife of beer and act synergistically with sulfur dioxide in wine.

Ascorbic acid is used to increase the volume of bread by assisting the formation of the gluten network. It is also added to foods to provide vitamin C as a specific nutrient. It is readily water soluble.

Sodium ascorbate and calcium ascorbate perform the same functions.

Limitations
In the EU, ascorbates are included in Part C Group 1of Regulation 1129/2011, additives permitted at *quantum satis*. Ascorbic acid has a slight acid taste and is insoluble in oils; sodium ascorbate has a very slight salty taste. As raw materials, ascorbates are gradually oxidised and should be kept in the dark in sealed containers. Ascorbic acid solutions have a low pH and, where this is a problem, the ascorbates should be used.

Typical Products
Bread, canned fruit and vegetables.

Number	Products
E 304	Fatty acid esters of ascorbic acid
	(i) ascorbyl palmitate
	(ii) ascorbyl stearate

Sources
The fatty acid esters of ascorbic acid are made by a two stage process involving ascorbic acid, sulfuric acid and the individual fatty acids.

Function in Food
The fatty acid esters of ascorbic acid are used to provide the antioxidant capacity of ascorbic acid to oils and fats.

Benefits
The esters are somewhat soluble in fats and oils. The palmitate is more common than the stearate. The reaction with the fatty acid does not affect the antioxidant capacity of ascorbic acid and it still has value to humans as the esters break down in the digestive tract, releasing ascorbic acid.

It is synergistic with dl-α-tocopherol, which is beneficial when stabilising oils with a natural tocopherol content.

Limitations
In the EU, the ascorbic acid esters are included in Group 1 Part C of Regulation 1129/2011, additives permitted at *quantum satis*. Because of their limited solubility in oils, they are best dissolved in hot oil before adding to other

ingredients, but the temperature needs to be controlled carefully since the palmitate ester, for example, decomposes at 113 °C. As raw materials, the ascorbyl esters are gradually oxidised and should be kept in the dark in sealed containers. The oxidation in solution is catalysed by metal ions, and the protective effect will be limited if metal ions are present in the final product.

Typical Products
Fat spreads, oils and fats.

Number	Product
E 306	Extracts of natural origin rich in tocopherols (natural vitamin E)
E 307	Synthetic α-tocopherol (synthetic vitamin E, dl-α-tocopherol)
E 308	Synthetic γ-tocopherol (synthetic vitamin E, dl-γ-tocopherol)
E 309	Synthetic δ-tocopherol (synthetic vitamin E, dl-δ-tocopherol)

Sources
Tocopherols may either be extracted from vegetable sources – such as oils from soya beans and sunflower seeds, nuts and grains, or produced by chemical synthesis. Although identical in terms of molecular composition, natural and synthetic tocopherols nevertheless exhibit fundamental differences, which are dependent on their origins. For their antioxidant effects in foods, these differences are unimportant, but the tocopherols of natural origin have been shown to be significantly more beneficial in human health and nutrition.
The natural tocopherols each contain only one of the eight possible steroisomers of the molecule, but the synthetic forms are always a mixture of all eight isomers.
The term "vitamin E" is often used as a general description for the four, closely related, naturally derived tocopherols, all showing some biological activity, of which the most active and potent is d-α-tocopherol. In the naturally derived material, as in the synthetic, the three companion compounds are the β, γ and δ isomers. The β isomer is present at only about 1% in natural extract and is ignored in the E number nomenclature. The tocopherols are clear, yellow oily liquids, which darken on exposure to light. The tocopherol-rich extract tends to be a darker colour than the individual tocopherols.

Function in Food
The tocopherols are antioxidants. They are fat soluble and are added to fats and oils to delay or prevent rancidity. It is well to remember that antioxidants cannot reverse or repair damage already done by oxidative processes; nor can they totally prevent it. Only by their presence before, or very soon after, the oxidative process begins can they significantly delay the onset of detectable rancidity in fats and oils, which is one of the major causes of the generation of off-flavours in food.
The antioxidant effect of tocopherols and the synthetic antioxidants such as butylated hydroxyanisole (BHA) and butylated hydroxytoluene (BHT) arises

from the presence of a hydroxyl group attached to an aromatic ring substituted with methyl groups. This molecular configuration permits donation of a hydrogen atom from the hydroxyl group to a fatty radical, thereby "quenching" it and stopping its catalytic effect in the degradation of oils and fats.

Benefits
The tocopherols also continue their antioxidant role after consumption.

Limitations
In the EU, the tocopherols are included in Group1 Part C in Regulation 1129/2011, additives permitted at *quantum satis*.

Typical Products
Margarine and low-fat spreads.

Number	Product
E 310	Propyl gallate
E 311	Octyl gallate
E 312	Dodecyl gallate

Sources
The gallates are white, odourless powders prepared by reaction of the appropriate alcohol with gallic acid. They have a slightly bitter taste.
Propyl gallate is also prepared from pods of the Tara tree (*Caesalpinea spinosa,* see also E 417) by extraction with propan-1-ol and subsequent purification.
Propyl gallate is the only one of the three gallates in commercial production.

Function in Food
The gallates are antioxidants. They are fat soluble and are added to fats and oils to delay or prevent rancidity.

Benefits
Propyl gallate is synergistic with other antioxidants, such as BHA and BHT. It is particularly effective with polyunsaturated fats. The antioxidant activity is maintained when the fat is blended with other ingredients in a final foodstuff. The longer chain length of the octyl and dodecyl gallates gives advantages over propyl gallate in terms of greater solubility in fats (and therefore less loss when the fats are emulsified in water) and greater stability.

Limitations
The gallates are regulated in the EU in Regulation 1129/2011. They are permitted in a range of fats and frying oils with individual limits, either alone or in combination with BHA and tertiary butyl hydroquinone (TBHQ).
The gallates are fat soluble but need to be dissolved in a small quantity of hot fat first before being diluted with the bulk of the fat.

Typical Products
Tallow, polyunsaturated oils.

Number	Product
E 315	Erythorbic acid
E 316	Sodium erythorbate

Sources
Erythorbic acid and sodium erythorbate are stereoisomers of ascorbic acid (E 300) and sodium ascorbate (E 301), respectively. Unlike ascorbates, they do not occur naturally, but are manufactured by a combination of fermentation and organic synthesis.

Function in Food
The erythorbates have the same antioxidant activity as the ascorbates, but with minimal vitamin activity. Therefore, where vitamin activity is not required, erythorbates are used as cost-effective general food antioxidants.

Benefits
Erythorbates are strong reducing (oxygen-accepting) agents, which gives rise to their antioxidant properties. Under many conditions, added erythorbates are preferentially oxidised in foods, thus preventing or minimising oxidative flavour and colour deterioration, and extending shelflife.

Limitations
Erythorbates are permitted in the EU in Regulation 1129/2011 in a limited range of fats and oils, in cured meat and certain fish products. The Scientific Committee for Food in the EU allocated erythorbic acid and sodium erythorbate an ADI of 0–6 mg/kg. JECFA allocated an ADI of "not specified" and, as a result, in many parts of the world, erythorbates are used as general food antioxidants at *quantum satis*. In the USA they have GRAS status when used as antioxidants in accordance with GMP.

Typical Products
Preserved meat and fats.

Number	Product
E 319	Tertiary butyl hydroquinone (TBHQ)

Sources
Tertiary butyl hydroquinone is a white, crystalline solid that is chemically synthesised from hydroquinone. It is soluble in ethanol, fats and oils but insoluble in water.

Function in Food
TBHQ is an antioxidant used to delay rancidity in fats and oils.

Benefits
Unlike propyl gallate, TBHQ does not cause discoloration in the presence of iron. It can be effective at very low dosages.

Limitations
TBHQ is readily soluble in fats and oils but, since such low levels are needed in foodstuffs, it is best to dissolve the required amount in a small amount of warm oil and then dilute with the bulk of the fat. In the EU, TBHQ is included in Regulation 1129/2011 where it is permitted in a range of products with individual limits, either alone or in combination with other antioxidants.

Typical Products
Fats and oils.

Number	Product
E 320	Butylated hydroxyanisole (BHA)

Sources
Butylated hydroxyanisole is a mixture of the 3-tert-butyl and 2-tert-butyl derivatives of 4-hydroxyanisole (also called 4-methoxyphenol). It is produced by chemical reaction of p-methoxyphenol and isobutene. Preparations usually consist mainly of the preferred 3-tert-butyl isomer.
BHA is available as a white or pale yellowish powder, large crystals or flakes with a waxy appearance and slight aromatic smell. It is soluble in fats, oils, alcohol and ether; but insoluble in water.

Function in Food
Butylated hydroxyanisole is an antioxidant and is added to delay or prevent rancidity in fats and oils in foodstuffs. It is insoluble in water and is best suited to foods with a high fat content. BHA is often used in combination with other antioxidants such as BHT to give a synergistic effect.

Benefits
BHA is stable to heat and mildly alkaline conditions, giving it a property of "carry-through" – a property that makes BHA particularly suitable for use in baked and fried foods.

Limitations
BHA is a very effective antioxidant for animal fats, but its effect is less marked in vegetable oils that are being stored at ambient temperatures. It has been allocated an ADI by JEFCA of 0–0.5 mg/kg body weight. Recommended usages rates of BHA are typically 100–200 mg/kg, based on the oil content. It is permitted in the EU in Regulation 1129/2011 in a range of fats and frying oils with individual limits, either alone or in combination with gallates or tertiary butyl hydroquinone (TBHQ).

Typical Products
Frying oils, animal fats

Number	Product
E 321	Butylated hydroxytoluene (BHT)

Sources
Butylated hydroxytoluene, 2,6-di-tert-butyl-4-methylphenol, is produced by the chemical reaction between p-cresol and isobutylene.
BHT is a white crystalline solid, either odourless or having a slight aromatic smell. It is soluble in alcohol and ether but insoluble in water.

Function in Food
Butylated hydroxytoluene is an antioxidant and is added to delay or prevent rancidity in fats and oils in foodstuffs. It is insoluble in water and is best suited to foods with a high fat content. BHT is often used in combination with other antioxidants, such as BHA, to give a synergistic effect.

Benefits
The antioxidant activity of BHT can be transferred to baked foodstuffs if it is used as an antioxidant in the shortenings used in their manufacture. The "carry-over" properties of BHT are not as good as those of BHA.

Limitations
BHT is more steam-volatile than BHA, and this makes it unsuitable for use on its own in frying oils, particularly where high-moisture foods are being fried. JECFA has allocated an ADI of 0–0.3 mg/kg body weight and EFSA an ADI of 0–0.25 mg/kg. Recommended usage rates of BHT are typically 100–200 mg/kg, based on the oil content. Within the EU, in Regulation 1129/2011, BHT is permitted in a limited number of foods with individual maxima in each case.

Typical Products
Tallow and fats

Number	Product
E 322	Lecithin

Sources
Lecithin is a mixture or fraction of phospholipids, obtained from animal or vegetable foodstuffs by a physical process. Lecithins are present in a range of foodstuffs including eggs, milk and marine sources but the major commercial products are derived from soya and, increasingly, sunflower. A number of different lecithins or lecithin fractions are available, including a more water-dispersible hydrolysed lecithin obtained by use of enzymes. The two main

products in commerce are standardized fluid and de-oiled powder lecithin. Speciality products include those with an enriched phosphatidylcholine content or from minor sources.

Function in Food

Lecithin is used to stabilise emulsions of oil and water.

Phospholipids are the active ingredients of lecithin and have a two-part molecular structure. One end of the molecule consists of fatty acids that are lipophilic (high affinity for fat/nonpolar phase) while the other end is a group containing phosphorus, *e.g.* phosphatidylcholine or phosphatidylinositol, which is hydrophilic (high affinity for water/polar phase). In emulsions, the phospholipids tend to dissolve in fat and disperse in water so that they are located at the interface between the two and can be used to stabilise both water-in-oil and oil-in-water emulsions.

In addition to emulsion stabilisation, lecithin is also used to moderate the flow (both viscosity and yield point) of chocolate during moulding and coating operations, to improve the wettability of powders in "instant" products, and to optimise the gluten network in baked goods.

Benefits

Lecithin allows the production of fine, stable emulsions with little aggregation or coalescence. It is used in chocolate manufacture to modify the flow characteristics of liquid chocolate for both blocks and coating. Lecithin is used on the surface of powders to improve "instant" properties. In bakery applications, lecithin is used to increase the extensibility of the gluten in bread making, and in batters to improve the overall distribution of ingredients in cakes and to assist the release of wafers from hot iron moulds.

Limitations

Lecithin has GRAS status in the USA and is included in Group 1 of Part C of EU Regulation 1129/2011, additives permitted at *quantum satis*.

Typical Products

Margarines, dressings, chocolate and confectionery, instant powders and bakery goods.

Number	Product
E 325	Sodium lactate
E 326	Potassium lactate

Sources

Sodium and potassium lactate are produced by neutralisation of lactic acid by sodium hydroxide or potassium hydroxide, respectively. They are available from both natural and synthetic sources.

Function in Food
The main functions of sodium and potassium lactate in food are in controlling spoilage and pathogenic bacteria, as flavouring and in pH regulation where they are used as buffer salts.

Benefits
Both sodium and potassium lactate can be used in pH-neutral food products, such as meat, poultry and fish. Many ingredients become effective only at lower pH, but both sodium and potassium lactates are effective in controlling both spoilage and pathogenic bacteria at neutral pH. They are used at levels of 2 to 4%.
Potassium lactate is one of the least bitter-tasting potassium salts available, and can be widely used in the food industry. Both lactates are used as buffer salts in confectionery products, cooking sauces and other savoury flavours.
Both lactates are used to control the fermentation of fermented products such as sausages, fermented dairy products and vegetable pickles.
The potassium salt is used in place of the sodium salt to contribute to the reduction of sodium in finished products.

Limitations
Both lactates have GRAS status in the USA and, within the EU, they are included in Part C Group 1 of Regulation 1129/2011, additives permitted at *quantum satis*.

Typical Products
Fresh meat products, sausages, ham.

Number   Product
E 327   Calcium lactate

Sources
Calcium lactate is produced by neutralisation of lactic acid by calcium hydroxide, chalk or lime. The lactic acid can be D/L-lactic acid or natural L-lactic acid. The natural form is about twice as soluble as the synthetic form.

Function in Food
Calcium lactate is used as a source of calcium; for nutrient fortification, reaction with pectins in fruit to improve texture, and coagulation of proteins.

Benefits
Calcium lactate, especially the L-form, is very soluble, and is in fact one of the few calcium sources that is soluble in low-pH environments (fruit juice, beverages, pickles, *etc.*) and neutral-pH environments (milk, diet food, infant food, *etc.*). Further, calcium L-lactate has a neutral flavour, is highly bio-available and is easily metabolised by the human body.

Much research has been carried out on the reduction of acrylamide in savoury products. It has been shown that the inclusion of calcium lactate in the formulation can reduce the level of acrylamide in savoury snacks.

Limitations
Calcium lactate has GRAS status in the USA and in the EU, is included in Part C Group 1 of Regulation 1129/2011, additives permitted at *quantum satis*. Within the EU, the use of calcium lactate in fortification of foods is regulated separately, under Regulation 1925/2006.

Typical Products
Soft drinks, fruit pastes, pickles, canned fruits.

Number	Product
E 330	Citric acid

Sources
Citric acid is a key intermediate in the human metabolic cycle. It occurs very widely in nature, notably in citrus fruits. It was first produced by extraction from lemon juice, but, since the 1920s, it has been made commercially by the large-scale fermentation of sugars using the mould *Aspergillus niger* or yeasts. After a series of purification steps and depending upon the temperature of crystallisation, either the monohydrate or the anhydrous form is obtained. Citrus acid is also produced as an aqueous solution, typically 50%w/w.

Function in Food
The primary functions of citric acid in food are as an acid, acidity regulator, antioxidant and sequestrant.
In dilute solution citric acid reduces the discoloration and spoilage of cut fruits, vegetables and shellfish. It helps presents rancidity in fats and aids the degumming of vegetable oils.
Citric acid is the most commonly used of all additives.

Benefits
The main characteristic of citric acid is its clean, tart taste, which is compatible with a very wide range of food flavours, both fruit and savoury. As well as providing flavour, its addition to a food formulation lowers pH, which inhibits microbial growth and spoilage. It has antioxidant properties, protecting sensitive flavours, and it is a powerful sequestering agent, binding metal ions that catalyse oxidation reactions leading to rancidity.
The monohydrate is the "traditional" form; nowadays, most food and beverage formulations are based on the most cost-effective anhydrous form or solution.

Limitations
Citric acid has been allocated an ADI of "not specified". Within the EU it is included in Part C Group 1 of Regulation 1129/2011, additives permitted at *quantum satis*. In practice its levels of use are limited by its strong flavour.

In foods with an acidic pH, which must be controlled accurately, one should use a buffered mixture of citric acid and citrate. Sodium citrate (E 331) is the most widely used, but potassium citrate (E 332) may be utilised in low-sodium foods. Citric acid is chemically stable in both its dry form and in solution. It is an irritant, so due care should be taken when handling.

Typical Products
Soft drinks, confectionery, preserves, soups and sauces.

Number	Product
E 331	Sodium citrates
	(i) monosodium citrate
	(ii) disodium citrate
	(iii) trisodium citrate

Sources
Citrates are found widely in nature.
The various sodium salts are produced industrially by either partially or completely neutralising citric acid with sodium hydroxide or carbonate. Trisodium citrate dihydrate, which is crystallised with minimum assay 99%, is the most commonly used form.

Function in Food
The sodium citrates are used primarily as acidity regulators, either in combination with citric acid or with acids naturally present in the food formulation. Trisodium citrate is used as an emulsifier in processed cheese and, in combination with ascorbate or erythorbate, it is effective as a cure accelerator in processed meat products.

Benefits
Trisodium citrate is effective and easy to use for pH control in food and beverage products. The appropriate combination of acid and salt can yield pH values from around 6 down to 2. In dry mixes, where moisture content must be minimised, a pH at the upper end of the range can be achieved with anhydrous trisodium citrate and at the lower end with anhydrous monosodium citrate.

Limitations
The sodium citrates have been allocated an ADI of "not specified". In the EU, in Regulation 1129/2011, they are included in Part C Group 1, additives permitted at *quantum satis*.

Typical Products
Trisodium citrate dihydrate is used in soft drinks, desserts, confectionery preserves.

The other sodium salts, such as anhydrous trisodium citrate and anhydrous monosodium citrate are used in dry foods and beverage formulations.

Number	Product
E 332	Potassium citrates
	(i) monopotassium citrate
	(ii) tripotassium citrate

Sources
The commercially available form of potassium citrate is the tripotassium salt in the monohydrate form. It is produced by neutralising citric acid, usually with potassium hydroxide, followed by crystallisation and drying.

Function in Food
Potassium citrate is used in foods as an acidity regulator and as a source of potassium ions.

Benefits
Potassium citrate has similar properties to sodium citrate, but it offers greater water solubility and can be used instead of the sodium salt in low-sodium foods. It is also a source of potassium ions in nutritional supplements.

Limitations
Potassium citrate has been allocated an ADI of "not specified" by JECFA. Within the EU it is included in Part C Group 1 of Regulation 1129/2011, additives permitted at *quantum satis*. The crystals are very hygroscopic and great care should be taken to prevent them from taking up moisture. When damp, potassium citrate will remain chemically stable but may cake hard. Potassium citrate has a diuretic effect.

Typical Products
Beverages and confectionery

Number	Product
E 333	Calcium citrates
	(i) monocalcium citrate
	(ii) dicalcium citrate
	(iii) tricalcium citrate

Sources
The commercially available form of calcium citrate is the tricalcium salt, formed by completely neutralising citric acid with calcium ions. It is available as a powder.

Function in Food
Tricalcium citrate is used as an acidity regulator and as a source of calcium ion.

Benefits
Calcium citrate is a physiologically acceptable source of calcium ion. It is effective in the formation of acid-based gels, with both alginates and pectin.

Limitations
The calcium citrates have been allocated an ADI of "not specified" by JECFA. In the EU, in Regulation 1129/2011, they are included in Group 1 Part C, additives permitted at *quantum satis*. Calcium citrate has minimal solubility in water, which decreases with increase in temperature. However, lowering of pH greatly increases solubility.

Typical Products
Dietary supplements and processed vegetables.

Number	Product
E 334	L(+)tartaric acid

Sources
Most of the tartaric acid in commerce is made from the acid potassium tartrate produced as a byproduct of the fermentation of grape juice into wine. It is also synthesised from malic acid.

Function in Food
Tartaric acid has two functions in foodstuffs; to provide a distinctive acid taste to finished products and, as part of baking powder, to react with carbonates to generate carbon dioxide.

Benefits
Tartaric acid has a different taste profile from citric acid, imparting less fresh and more sour notes to products. Naturally, it blends better with grape flavours than with citrus.
Tartaric acid is available as a powder. It is the most water soluble of the solid acids and is used in baking powders.
It is used as a chelating agent for metal ions naturally present in products, and hence acts as a synergist for antioxidants.

Limitations
Tartaric acid has been allocated an ADI of 0–30 mg/kg body weight (combined for the acid and its salts). In the EU, the acid is included in Group 1 Part C of Regulation 1129/2011, additives permitted at *quantum satis*.

Typical Products
Baking powder, biscuits and jams.

Number	Product
E 335	Sodium tartrates
	(i) monosodium tartrate
	(ii) disodium tartrate
E 336	Potassium tartrates
	(i) monopotassium tartrate
	(ii) dipotassium tartrate
E 337	Sodium potassium tartrate

Sources
Monopotassium tartrate is formed as a byproduct of the fermentation of grape juice. The dipotassium salt is produced by reaction of this with potassium hydroxide.
Sodium tartrates are produced from commercial tartaric acid.
Sodium potassium tartrate is also known as "Rochelle salt", which occurs as a crystalline deposit during the production of wine.

Function in Food
Monopotassium tartrate is also known as "cream of tartar". It is used as a source of acidity in baking powders. Monosodium tartrate would be equally effective but is not so readily available.
The tartrates are also used as buffers and taste modifiers in products containing tartaric acid.

Benefits
Monopotassium tartrate is one of the fastest reacting acidulants used in baking powder. It also softens the dough but does not weaken it so much that the evolved gas is lost.

Limitations
The tartrates are included in the EU in Part C Group 1 of Regulation 1129/2011, additives permitted at *quantum satis*.
In the past, tartrates were used as emulsifying salts in the production of processed cheese, but they have been replaced because calcium tartrate crystals tended to form during the process, and these crystals gave the appearance of fragments of glass, which alarmed consumers.

Typical Products
Monopotassium tartrate is used in baking powder.

Number   Product
E 338   Phosphoric acid

Sources
Phosphoric acid is manufactured commercially by the addition of sulfuric acid
to phosphate rock, followed by additional steps to remove impurities. An
alternative approach is to burn phosphate rock in an electric furnace to form
elementary phosphorus that is then burned in air to phosphorus pentoxide. This
is then hydrated to form phosphoric acid that is purified with hydrogen sulfide.

Function in Food
Phosphoric acid is used as an acidulant in soft drinks, jams, cheese and beer. It
is the only inorganic acid used extensively as a food acid. It is also used as a
setting aid and a sequestrant, and in sugar refining.

Benefits
Phosphoric acid is one of the cheapest and strongest food-grade acids available.
Its sharp acid flavour particularly complements the dry character of cola
drinks – better than citric or tartaric acids. The low pH generated by phosphoric acid is synergistic with other preservatives.

Limitations
In the EU phosphoric acid is permitted, together with phosphates, in Regulation 1129/2011 in a range of products with individual limits in each case.

Typical Products
Soft drinks, jam.

Number   Product
E 339   Sodium phosphates
                                                  (i) monosodium phosphate
                                                  (ii) disodium phosphate
                                                  (iii) trisodium phosphate
E 340   Potassium phosphates
                                                  (i) monopotassium phosphate
                                                  (ii) dipotassium phosphate
                                                  (iii) tripotassium phosphate

Sources
The phosphates are prepared by reaction of metal hydroxides (E 524 and E 525)
with phosphoric acid (E 338) under conditions controlled to maximise the yield
of the required product. The crystalline phosphates are separated and dried.
The products are available both as hydrated crystals and as dehydrated
powders. The hydrated form can lose water or cake on storage.
The two sets of products are considered together because they are largely
interchangeable, although the potassium salts are generally more soluble than
their sodium equivalents.

Function in Food

Monometalphosphates

The monometalphosphates are acidic and can be used as acidulants in raising agents. However, the calcium salt is more commonly used for this purpose (see E 341(i) and (ii)).

The sodium and potassium salts are used as chelating agents, buffering agents and occasionally as emulsifying salts in processed cheese products. They are also used as a component of mixtures with other phosphates for protein binding in meat products.

Dimetalphosphates

The dimetalphosphates are used for their ability to enhance water binding in meat and dairy products, preventing water loss and shrinkage during cooking and storage. They are used to stabilise milk products, such as evaporated milk where they prevent protein coagulation and gelling on storage. They are also used to increase the rate of gelling in instant puddings and cheesecakes. They are powerful sequestrants of calcium in water and are used as such to prevent flocculation of milk proteins during rehydration of milk-based powders in hard water. The dimetalphosphates are the most important emulsifying salts in processed cheese, because they provide the required body and melting performance without fat separation. They are often used in combination with the trimetal phosphates and occasionally with monometal phosphates.

Trimetalphosphates

Only the trisodium phosphate is of any commercial significance. It is alkaline and is used as a buffering agent and texturiser in meat and cheese products. It is also used to increase the speed of cooking of peas, beans and cereals. The main use is in industrial detergents and toothpastes but, since it is strongly alkaline, it is also used to reduce the microbial load on animal carcasses.

Limitations

In the EU the phosphates are included in Regulation 1129/2011 where they are permitted in a wide range of products with individual maxima in each case.

Typical Products

Processed cheese, cooked ham, desserts, evaporated milk.

E 341   Calcium phosphates
                                         (i) monocalcium phosphate
                                         (ii) dicalcium phosphate
                                         (iii) tricalcium phosphate

Sources

The calcium phosphates are manufactured by the reaction of calcium hydroxide (hydrated lime) (E 526) and phosphoric acid (E 338) under conditions controlled to maximise the yield of the required product.

Function in Food

Monocalcium phosphate

Monocalcium phosphate is used as a raising agent when rapid reaction with sodium bicarbonate is required. Unlike the dicalcium salt, the reaction commences as soon as the phosphate is added to the cake batter. Recently, however, mixtures with other phosphates have been developed, which allow a slower rate of reaction, and a more even release of gas.

Monocalcium phosphate is also added to flour to reduce the risk of growth of the bacteria that lead to the spoilage condition known as "rope". It is used as a source of calcium to improve the structure obtained from low-gluten flours, to increase the rate of gelling of some milk-based desserts, and to increase the firmness of canned vegetables such as carrots and tomatoes.

Dicalcium phosphate

Dicalcium phosphate is available in both dehydrated and anhydrous forms. The dehydrate is used as a raising agent in combination with other phosphates and sodium bicarbonate. Dicalcium phosphate is practically insoluble in water and does not react until the cake is heated to about 60 °C, when it dehydrates and decomposes. It is only mildly acid, having a neutralising value half that of disodium phosphate. It is used in products that require a baking time in excess of 30 min and in combination with faster-acting raising agents when it provides last-minute expansion of the cake batter just before the batter sets.

Tricalcium phosphate

Tricalcium phosphate is used as a free-flow agent in the powdered materials such as icing sugar and powders for instant drinks. Being a fine powder it is used to coat the surfaces of other materials to improve the flowability of the mix and reduce the propensity to form clumps.

Limitations

In the EU, these phosphates are included in Regulation 1129/2011 where they are permitted in a range of products with individual limits in each case.

Typical Products

Monocalcium phosphate is used in cakes, canned fruit and milk desserts.
Dicalcium phosphate is used in cakes.
Tricalcium phosphate is used in powders for making drinks.

Number	Product
E 343	Magnesium phosphates
	(i) monomagnesium phosphate
	(ii) dimagnesium phosphate

Sources

The phosphates are prepared by the reaction of magnesium oxide (E 530) with phosphoric acid (E 338) under conditions controlled to maximise the yield of the required product. The crystalline phosphates are separated and dried.

The products are available both as hydrated crystals and dehydrated powders. The hydrated form can lose water or cake on storage.

Function in Food
Magnesium phosphates are used as acidulants in raising agents in dough.

Benefits
The magnesium phosphates react slowly and are used to stabilise doughs that will be held for some time before baking. They fulfil a similar function to sodium aluminium phosphate.

Limitations
In the EU, magnesium phosphates are included with other phosphates in Regulation 1129/2011, where they are permitted in a range of products with individual limits in each case.

Typical Products
Bakery goods.

Number	Product
E 350	Sodium malates
	(i) Sodium malate
	(ii) Sodium hydrogen malate
E 351	Potassium malate
E 352	Calcium malates
	(i) Calcium malate
	(ii) Calcium hydrogen malate

Sources
The malates are made by reacting malic acid with the appropriate hydroxide or carbonate.

Function in Food
The malates are acidity regulators to buffer and modify the acid taste of products containing malic acid.

Benefits
The malates complement the flavours of products, such as those with apple flavours, acidified with malic acid. The sodium salt is more common than the potassium, which would only be used if the sodium content of the product needed to be restricted.

Limitations
The malates are included in Group 1 Part C of Regulation 1129/2011 in the EU, additives permitted at *quantum satis*.

Typical Products
Jam.

Number	Product
E 353	Metatartaric acid

Sources
Metatartaric acid is manufactured from glucose. It is also known as glucaric acid.

Function in Food
Metatartaric acid is used as a sequestrant to prevent deposition of cream of tartar (potassium hydrogen tartrate) and calcium tartrate in wine during storage.

Limitations
In the EU, in Regulation 1129/2011, metatartaric acid is only permitted in made wine and then only up to 100 mg/l.
It is deliquescent and should be kept in tightly closed packages.

Number	Product
E 354	Calcium tartrate

Sources
Calcium tartrate is prepared as a byproduct of the wine industry.

Function in Food
Calcium tartrate is used as a buffer and as a preservative.

Limitations
Calcium tartrate is included in Part C Group 1 of Regulation 1129/2011 in the EU, additives permitted at *quantum satis*.

Typical Products
None known.

Number	Product
E 355	Adipic acid

Sources
Adipic acid is produced by the oxidation of cyclohexane.

Function in Food
Adipic acid is used to provide an acid taste.

Benefits
Adipic acid is used to provide an acid taste with a more lingering flavour profile than citric acid, which works well with some noncitrus fruit products.
It is practically nonhygroscopic.

Limitations
In Regulation 1129/2011 in the EU, adipic acid is only permitted in fillings for bakery products, dessert mixes and powders for home preparation of drinks, with individual limits specified for each usage. The limits are maxima for any single or combined use of E355, E356 and E357.
Adipic acid has been allocated an ADI of 0–5 mg/kg body weight. This covers adipic acid alone or in combination with the sodium or potassium salts of adipic acid.

Typical Products
Individual pies with fruit filling.

Number	Product
E 356	Sodium adipate
E 357	Potassium adipate

Sources
The adipates are made by reaction of adipic acid with the appropriate hydroxide or carbonate.

Function in Food
The adipates are used to buffer the acidity and modify the acid taste of formulations containing adipic acid.

Benefits
The adipates complement the flavour of products acidified with adipic acid. The sodium salt is more common than the potassium, which would only be used if the sodium content of the product needed to be restricted.

Limitations
In Regulation 1129/2011 in the EU, the adipates are only permitted in fillings for bakery products, dessert mixes and powders for home preparation of drinks, with individual limits specified for each usage. The limits are maxima for any single or combined use of E355, E356 and E357.
The adipates have been allocated an ADI of 0–5 mg/kg body weight. This covers the adipates alone or in combination with adipic acid.

Typical Products
Individual pies with fruit filling.

Number	Product
E 363	Succinic acid

Sources
Succinic acid occurs naturally in a wide range of vegetables but is manufactured from acetic, fumaric or maleic acids.

Function in Food
Succinic acid is used to provide a distinctive acid taste.

Benefits
Succinic acid is water soluble but not hygroscopic, which makes it useful in powdered products.

Limitations
In the EU succinic acid is included in Regulation 1129/2011 where it is permitted in desserts, soups and broths and in powders for home preparation of drinks, each with maximum permitted levels. The acid has a pronounced aftertaste and dissolves only slowly in water.

Typical Products
None known.

Number	Product
E 380	Triammonium citrate

Sources
Triammonium citrate is the final product of the reaction between citric acid (E 330) and ammonium hydroxide (E 527). It is a white, water-soluble powder.

Function in Food
Triammonium citrate is little used in the food industry. Its only applications are as a yeast food and a chelating agent.

Limitations
Triammonium citrate is included in Part C Group 1 of Regulation 1129/2011 in the EU, additives permitted at *quantum satis.*

Number	Product
E 385	Calcium disodium EDTA

Sources
EDTA is ethylene diamine tetracetic acid.
Calcium disodium EDTA is the mixed salt of EDTA made by reacting the acid with a mixture of calcium and sodium hydroxides. EDTA itself is made by a multistage process starting from ethylene glycol (1,2 dihydroxy ethane).

Function in Food
Calcium disodium EDTA is a sequestrant, both binding metal ions and exchanging its calcium for metal ions.

Benefits
Calcium disodium EDTA is used to sequester small quantities of metal ions present in raw materials or process water. These metals tend to catalyse degradation reactions such as those leading to rancidity, and their removal increases the stability of products during storage and extends shelflife. By a similar mechanism it stabilises vitamin C and oil soluble vitamins.
It is used in spreadable fats as a synergist for the antioxidant vitamins, having an advantage over citric acid or polyphosphate in that it imparts no flavour. The salt is used because it is more stable than the acid.

Limitations
In the EU, calcium disodium EDTA is only permitted in Regulation 1129/2011 in a number of canned and bottled products, in spreadable fats and in emulsified sauces with individual maxima specified in each case.
Calcium disodium EDTA has been allocated an ADI of 2.5 mg/kg body weight by JECFA.

Typical Products
Catering sauces and salad dressings.

Number
E 392

Product
Extracts of rosemary

Sources
Rosemary, *Rosmarinus officinalis*, is an evergreen shrub native to the Mediterranean region and the Caucasus. The leaves are extracted with one of a number of organic solvents or supercritical carbon dioxide and subsequently deodorised. The extract is naturally oil soluble and is standardised with vegetable oil. The active ingredients are carnosic acid and carnosol and the material is specified in terms of the total content of these two components.

Function in Food
The extract is used as an antioxidant to protect the colour and flavour of meat products during their shelflife.

Benefits
Rosemary extract is an effective antioxidant for meat and oil products, with the advantage of being derived from a natural raw material.

Limitations
In the EU, rosemary extract is included in Regulation 1129/2011 where it is permitted in a number of fats and meat products with individual limits in each case.

Typical Products
Meat products and vegetable oils.

Number	Product
E 400	Alginic acid

Sources

Alginates are the principal structural polysaccharide components of brown seaweeds (just as cellulose is the principal carbohydrate in land plants). The commercial product is extracted from a wide range of brown seaweed species, *e.g. Ascophyllum* from the North Atlantic, *Macrocystis* from California and Mexico, *Lessonia* from South America, *Durvilea* from Australia, *Ecklonia* from South Africa and *Laminaria* from various northern hemisphere oceans. In general, seaweed harvested for alginate manufacture is material growing naturally, but there is some cultivation in China.

Alginate is present in seaweed as a mixed salt of sodium, potassium, calcium and magnesium. Extraction involves ion exchange in an alkaline medium followed by precipitation, purification and recovery of the alginic acid. Alginic acid is a copolymer of mannuronic acid and guluronic acid – two natural anionic sugars. The monomer composition and sequence vary, mainly as a consequence of the seaweed raw material.

Alginate can also be produced by microbial fermentation, but economics and the need for separate regulatory approval restrict this to a laboratory curiosity at the present time.

Function in Food

Alginic acid swells in water, but does not dissolve, and its main applications are in pharmaceutical tablets. Its swelling ability makes it a useful tablet disintegrant, and it is used in antacid tablets as a raft former for stomach disorders. In the food industry, it is rarely added directly to food compositions. However, it is produced *in situ* when sodium alginate (see E401) is used in acidic foodstuffs. In such situations it will form a gel, skin or fibre as a result of its insolubility in water. Alginic acid is also used in some formulated alginate products for stabilising ice cream and whipped dairy cream. In this case, the alginic acid is converted to sodium alginate *in situ* to provide the stabilisation.

Benefits

Alginic acid and alginates are not absorbed by the human body so are considered a low-calorie ingredient and possibly a source of dietary fibre. They are efficient water binders.

Limitations

In the EU, alginic acid is included in Part C Group 1 of Regulation 1129/2011, additives permitted at *quantum satis*. It is not permitted for use in jelly minicups.

It is insoluble in water and therefore rarely used directly as a stabiliser or gelling agent.

Typical Products
Ice cream and whipped cream.

Number	Product
E 401	Sodium alginate
E 402	Potassium alginate
E 403	Ammonium alginate
E 404	Calcium alginate

Sources
Alginates (see also alginic acid E400) are the principal structural components of brown seaweeds. They are present in seaweed as a mixed salt of sodium, potassium, calcium and magnesium. Extraction involves ion exchange in an alkaline medium followed by precipitation, purification and conversion to the appropriate salt.

Function in Food
The sodium, potassium and ammonium salts are cold-water soluble and are used interchangeably, but the calcium salt is insoluble. The salts are used for thickening, gelling, stabilising, film-forming and controlled-release applications.
Alginates are copolymers of mannuronic and guluronic acids and the monomer composition and sequence vary as a consequence of the seaweed raw material. In general, high guluronic acid alginates are used for gelling applications and the high mannuronic acid types for thickening and stabilising.
The soluble salts form viscous solutions in hot and cold water, and form gels by controlled reaction with calcium. The free calcium content of milk prevents the soluble alginates from dissolving directly in cold milk. This is overcome by the use of calcium sequestering agents or by dissolving it in milk at, or just below, its boiling point. When a soluble alginate is used as a suspending agent, small amounts of available calcium are beneficial. Any soluble calcium will increase the pseudoplastic nature (shear dependency) of the alginate solution. At rest, suspended solids, or oil droplets will be stabilised, but the liquid will still flow freely when sheared. Higher concentrations of calcium will produce a thixotropic system (shear-reversible gel) and higher concentrations still will produce a thermostable gel (*i.e.* it will not melt).
Alginate gels can be internally set, where the gelling ingredients are mixed in with the alginate. Internally set gels are formulated to set within a prede-termined time and need to be completely filled into their final container within this time. The careful formulation of partially soluble calcium salts and sequestrants into the product allows the setting time to be varied to fit production needs. Typically, calcium salts such as calcium sulfate and calcium phosphate, and sequestrants such as phosphates and citrates are used for this purpose. Externally set alginate gels rely on the diffusion of readily soluble

calcium salts (*e.g.* calcium chloride, calcium lactate) into food containing an alginate solution. If such a food is extruded into a setting bath containing calcium chloride, a skin of calcium alginate forms instantaneously. This gives the food a structurally robust form and shape. Further calcium diffusion into the centre of the food gels the alginate throughout. In frozen products, *e.g.* ice cream, sodium alginate prevents ice crystal and fat clump growth during melt/freeze cycles by restricting water mobility.

Benefits
Alginates are very efficient water binders and this leads to their use as thickeners, where low levels give high viscosities; as gelling agents; and in solid foods to prevent water loss, syneresis and phase separation. The cold solubility and the ability to make gels without the use of heat differentiates alginates from other hydrocolloids, *e.g.* gelatine, agar, carrageenan and locust bean gum, which all require high-temperature processes. This makes alginates particularly useful when used with heat-sensitive ingredients like flavours and in applications for safe, convenient domestic use such as instant mousse mixes and cheesecakes. The ability of alginates to form gels, skins and fibres makes them particularly useful for making restructured foods, for example onion rings and pet food chunks.
Alginates are not absorbed by the human body, so that they are a low-calorie ingredient.

Limitations
As with all hydrocolloids, care needs to be exercised in dissolving alginates. Careless addition leads to clumping, where the outside of the powder hydrates quickly, preventing powder inside the clump from dissolving. The use of appropriate mixing equipment and careful addition of the powder will avoid clumping. Alternatively, the soluble alginates can be dry blended, *e.g.* with sugar, or wetted with a nonsolvent oil or alcohol, prior to addition to water. This will allow each alginate particle to hydrate separately.
The soluble alginates will not dissolve directly in cold milk and other high calcium environments. Sequestrants are normally used to overcome this. Similarly, the soluble alginates will not hydrate in highly acidic environments (pH < 4–5 depending on grade). Soluble alginates can be used in foods with a pH as low as 3.5 but below this alginic acid precipitates out and in these conditions propylene glycol alginate should be considered as an alternative. Gels, once formed, are not thermally reversible.
In the EU, alginates are included in Part C Group 1 of Regulation 1129/2011, additives permitted at *quantum satis*. They are not permitted in jelly minicups.

Typical Products
Sauces, salad dressings, desserts, fruit preparations, ice cream and water ices, onion rings, low-fat spreads and bakery filling creams. Ammonium alginate is particularly used for icings and frostings.

Number	Product
E 405	Propylene glycol alginate (PGA)
	Propane 1,2 diol alginate

Sources

Propylene glycol alginate is made by esterification of alginic acid (E 400). This varies in composition as a result of the source of the raw material, its degree of esterification and the percentage of free and neutralised carboxylic acid groups in the molecule.

Function in Food

Propylene glycol alginate (PGA) is a thickener, suspending agent and stabiliser. It forms viscous solutions in hot and cold water. It may be used in many of the same applications as the soluble alginates but has the advantage of being more compatible with more acidic foods and foods with a significant calcium content. The higher the degree of esterification, the better the compatibility. However, PGA does not form gels, films or fibres with calcium. The residual sensitivity of low- or medium-esterified PGA to calcium can enhance its rheology and provide superior suspending and stabilisation properties, but does prevent it from dissolving in milk below the boiling point. Grades with a high degree of esterification interact with proteins and are used to stabilise beer foam, meringues and noodles.

Benefits

Propylene glycol alginate can be used in foods with pH as low as 3 and it is less sensitive to calcium than the other alginates. It is used in salad dressings to stabilise the oil-in-vinegar emulsion and in fruit drinks to prevent separation of pulp and flavour oils. It is a very efficient water binder, so low levels give high viscosities. It also interacts with proteins and is particularly useful with heat-sensitive systems.

Limitations

Care has to be taken in dissolving PGA; it hydrates very quickly leaving dry powder inside lumps. The use of appropriate mixing equipment and careful addition of the powder can solve this problem. Alternatively, the alginate can be dry mixed with a granular material such as sugar or wetted with a nonsolvent oil or alcohol before mixing with water. PGA does not dissolve in cold milk or other high calcium environments and, when using in acidic foods, it is better to dissolve the PGA in a neutral medium before adding acid. It is unstable at alkaline pH and if protein reactivity at alkaline pH is used in the application, the food product needs to be neutralised quickly after the reaction has occurred.

Within the EU, propylene glycol alginate is included in Regulation 1129/2011, where it is permitted in a range of foodstuffs with individual maxima in each case.

Typical Products
Salad dressings, meringues, ice cream, noodles, dairy desserts and beer.

Number	Product
E 406	Agar

Sources
Agar is obtained from red seaweeds of the *Gelidium* and *Gracilaria* species collected from the coasts of Japan, Korea, Chile, Spain, Portugal, Morocco and Indonesia. The agar is extracted using hot, dilute alkali. The solution is cooled to form a very firm brittle gel, which is frozen to disrupt the gel structure. When the gel is thawed, impurities dissolved in the water can be expelled using high pressure and the gel dried and ground to produce powdered agar. Very small amounts of "natural" strip agar are made from *Gelidium* seaweeds. Solutions are cast in moulds and the gels are frozen before pressing and drying to give strip agar used in traditional Oriental foods.

Function in Food
Agar forms thermally reversible, firm, brittle gels. These are generated by hydrogen bonds between adjacent chains of repeating units of galactose and 3,6 anhydro galactose. This gel structure is not affected by salts or proteins. The gel hysteresis, or difference between melting and setting points, is much greater with agar than with other gelling agents.
Agar is used to gel fermented dairy products in Europe, but by far the largest volume of agar continues to be used in Asia for traditional dishes of Tokoroten noodles, Mitsumame and Red Bean jelly and in fruit-flavoured water dessert jellies.

Benefits
Agar gels are completely reversible and may be melted and reset without any loss of gel strength. The gels have a characteristic firm brittle texture. Enhanced rupture strength and a more elastic texture are obtained by adding up to 20% locust bean gum, with a maximum synergy at 10% locust bean gum, to gels of *Gelidium* agar.

Limitations
Within the EU, agar is included in Part C Group 1 of Regulation 1129/2011, additives permitted at *quantum satis*, but it is not permitted in jelly minicups. In the USA, agar has GRAS status.
The tannic acid found in some fruits, such as quince and some varieties of apples and plums, can inhibit the gelation of agar.

Typical Products
Jams and marmalades, toppings and fillings for bakery products such as doughnut glaze. Gelled meats and confectionery jellies. Other applications are largely confined to specific cultural areas of the world.

Number	Product
E 407	Carrageenan

Sources
Carrageenans are extracted from red seaweeds of the class *Rhodophyceae*. Although some seaweed raw material is gathered from the shores, most is now farmed in areas such as the coast of the Philippines, Indonesia and east Africa. Carrageenan polymer chains are based on galactose and anhydrogalactose with varying amounts of natural sulfation.

Function in Food
Carrageenans are used as gelling agents, thickening agents and stabilisers. In dairy products, carrageenans are used to form gels with a range of textures, to thicken milk drinks and to stabilise neutral pH dairy products. Carrageenan is used to form water jellies, frequently in combination with locust bean gum. The water-gelling properties are widely used in cooked meat products to bind water, especially in cold-eating poultry and pork products. Different seaweed types give different carrageenan types on extraction, which have the designations kappa, iota and lambda. These three idealised types have differences in chemical structure which lead to differences in gel texture, with kappa types the strongest and most brittle, iota giving soft gels and lambda types being nongelling. Commercial products frequently are blends of more than one carrageenan type to produce the required textures.

Benefits
Carrageenans can interact with the casein protein in dairy products. This allows carrageenans to produce an equivalent effect in dairy products at lower use levels than most other food gums. The processes used in carrageenan extraction produce gels of good clarity, which is highly desirable in water jellies. Different gel textures are produced by the different seaweed raw materials and so a wide range of textures can be produced in many of the application fields. Carrageenan also has a synergy with locust bean gum, and this is used to extend the range of textures.

Limitations
Carageenan in solution is not stable to a combination of high temperature and low pH since this will degrade the polymer chain. Carrageenan solutions must therefore be subjected to minimal processing at pH levels below 4.0. Carrageenan is permitted in the EU in Group 1 of part C of Regulation 1129/2011, additives permitted at *quantum satis* but, like other gelling agents, it is not permitted in jelly minicups.

Typical Products
Dairy desserts, powder mixes for dairy desserts, milk drinks, creams and toppings and ice creams; water jellies and powder mixes for water jelly desserts; hams and cold-eating poultry products; glazes for bakery wares.

Number   Product
E 407a   Processed euchema seaweed (PES)

Sources
This additive is obtained by aqueous alkaline extraction of the red seaweed types *Euchema cottonii* and *Euchema spinosum* followed by washing, drying and milling.

Function in Food
Processed euchema seaweed (PES) is used as a gelling agent and water-binding agent, and as a thickener and stabiliser. It is used in hams and cold-eating cooked poultry products to bind water and to increase yields. It is also used to stabilise ice cream, to thicken flavoured milk drinks, to stabilise cocoa powder in chocolate milks and to gel dairy desserts.

Benefits
The simpler production process for PES allows a lower-cost product than is possible with carrageenan, and PES can be partially or totally substituted for carrageenan in a number of uses, especially in meat products. Substantial yield increases can be obtained in cold-eating cooked meat products by replacing carrageenan with PES.

Limitations
PES solutions are not stable to combinations of high temperature and low pH, and must therefore be subjected to minimal heat processing at pH values under 4.0. Insoluble cellulosic components in PES produce cloudy solutions that are unsatisfactory for many water jelly applications
In the EU, PES is included in Group 1 of Part C of Regulation 1129/2011, additives permitted at *quantum satis*, but it is not permitted in jelly minicups.

Typical Products
Hams, poultry roll and chocolate milk.

Number   Product
E 410   Locust bean gum

Sources
Locust bean gum (LBG) is the ground endosperm of the seed of the locust bean (carob) tree, *Ceratonia siliqua*, which grows wild in countries bordering the Mediterranean Sea. The principal producers are found in Spain, Morocco and Greece. The main component of the white powder (*ca.* 80%) is a high molecular weight linear polysaccharide (galactomannan) with a mannan backbone chain carrying single galactose residues. The distribution of these galactose sugars along the chain is not known, but statistically there are approximately four

mannose sugars present in the molecule for every galactose moiety. In addition to the native gum, LBG is also available in alcohol-washed and alcohol-precipitated qualities. This process removes much of the protein and other components from the gum, which then gives clear transparent solutions. Cold-soluble pregelatinised forms of LBG are also commercially available.

Function in Food
Locust bean gum is an efficient thickening and gelling agent. The powder partially hydrates in cold water, but the full viscosity can be obtained only by heating the solution to at least 85 °C. LBG forms thermoreversible gels when mixed with xanthan, ideally in the ratio 1:1. LBG also interacts synergistically with kappa-carrageenans to increase the strength and elasticity of the gels.

Benefits
LBG is used as a thickening agent in hot-prepared fabricated foods. It is widely used in combination with xanthan to prepare elastic gels, which, in comparison with other polysaccharide gelling systems (alginates, carageenans, pectins), are insensitive to the presence of common cations. It also forms synergistic mixtures with guar gum.
It is more resistant to shear than starch. It is nondigestible and may be classified as soluble fibre. In appropriate dosages, it is known to increase intestinal tract motility and reduce blood serum cholesterol levels.

Limitations
Locust bean gum is included in the EU in Group 1 of Part C of Regulation 1129/2011, additives permitted at *quantum satis*, but is not permitted in jelly minicups. Because it readily absorbs water and swells, it should not be ingested as a dry powder, and like some other gelling agents, is not permitted in dehydrated foods intended to rehydrate on ingestion. Isolated reports have appeared that indicate that the protein in LBG may act as an allergen. The incidence appears to be no higher than that associated with any other natural protein.

Typical Products
Ice cream and hot-prepared sauces, soups, ketchups and mayonnaises. It is often found together with xanthan as the gelling system in dressings, desserts and mousses.

Number	Product
E 412	Guar gum

Sources
Guar gum is the ground endosperm of the seeds of the guar plant (*Cyamopsis tetragonolobus*) which is cultivated in the arid regions of north-west India

(Rahjastan) and Pakistan. The main component (*ca.* 80%) is a galactomannan with a backbone of mannose to which are attached single galactose residues. The distribution of the galactose along the mannan chain is not known, but statistically there is approximately one galactose residue for every two mannose sugars. The typical molecular weights exceed 10^6 Daltons but depolymerised grades that show lower viscosity are also commercially available. The powder can be steam treated to remove much of the characteristic "beany" flavour. Guar from which some of the galactose residues have been enzymically removed so that it mimics the behaviour of locust bean gum towards xanthan and kappa-carrageenan is also on sale.

Function in Food
Guar gum is an efficient thickening agent. It dissolves almost completely in cold water to give opalescent pseudoplastic solutions, and shows a synergistic increase in viscosity when mixed with xanthan. By virtue of its size, guar can cause phase separation with other thermodynamically incompatible solutes. This effect has been exploited in the formulations of fat-reduced or fatless spreads.

Benefits
The pronounced pseudoplastic flow properties of guar solution are ideal for delaying sedimentation of solids or creaming of fats. They ensure that, at low shear forces, an effective viscosity is present without making the product unpalatable.
Guar has an advantage over starch in that it is more resistant to shear. It forms synergistic mixtures with locust bean and xanthan gums.
It is nondigestible and may be classified as a soluble fibre. In appropriate dosages it is known to increase intestinal tract motility and reduce blood serum cholesterol.

Limitations
Within the EU, guar gum is included in Group 1 of Part C of Regulation 1129/2011, additives permitted at *quantum satis*, although it is not permitted in jelly min-cups. Because it readily absorbs water and swells, it should not be ingested as a dry powder and is not permitted in dehydrated foods intended to rehydrate on ingestion. Isolated reports have appeared that indicate that the protein in guar gum may act as an allergen. The incidence appears to be no higher than that associated with any other protein-containing food. Following concerns over the treatment of the guar plants with chemical sprays, the EU issued Regulation 258/2010 which imposed special conditions on the imports of guar gum originating in or consigned from India due to the risk of contamination by pentachlorophenol and dioxins.

Typical Products
Ice cream, drinks, sauces, ketchups and mayonnaises, cold-prepared deep-frozen foods and as a flour additive in the bakery industry.

Number	Product
E 413	Tragacanth

Sources

Tragacanth is a natural gum exudate from wounds of shrubs of the species *Astragalus,* mainly *A. microcephalus* and *A. gummifer,* which grow in arid areas of Iran and Turkey. Incisions are made in the lower stem and roots of the shrub and the gum is exuded as thin white ribbons or larger off-white flakes. These are allowed to dry, collected, sorted by colour and milled to a fine powder. Traces of bark and foreign matter are removed before and during the milling process. A wide range of viscosity grades is available, with whiter ribbon grades generally possessing the highest viscosity and the flake form of tragacanth having the best emulsifying properties. Heat-treated variants with lower total viable counts (TVC) are also available. Tagacanth is a complex high molecular weight branched polysaccharide consisting of two main fractions. The major fraction (known as bassorin or tragacanthic acid) swells in water and the second fraction (tragacanthin) is water soluble. Bassorin has a 1-4 linked D-galactose backbone substituted by D-xylose or side chains of D-xylose with L-fructose or D-galactose. Bassorin occurs as a mixed calcium, magnesium and potassium salt. Tragacanthin is a neutral arabinogalactan with a 1-6 and 1-3 linked D-galactose backbone substituted with arabinose side chains. A proportion of protein (1–4%) is present in tragacanth and may be involved in its emulsifying properties.

Function in Food

Tragacanth is used as a cold-soluble thickener, stabiliser, suspending agent and emulsifier. It is also used as a processing aid in lozenge production and as a plasticiser in icings. Tragacanth can be used as a fat replacer in emulsion products.

Benefits

Tragacanth is an extremely effective thickener, giving high viscosity at low concentration. It is unusual in that it possesses both thickening and emulsifying properties; it will thicken and stabilise food emulsions and is particularly effective in pourable emulsions. The excellent acid stability of tragacanth has resulted in its widespread use in dressings. It is also resistant to hydrolysis by food enzymes. It possesses suspending properties and has a creamy mouthfeel with neutral flavour. In contrast to stabilisers such as xanthan gum it does not develop "stringy" rheology in high-solids systems.

Tragacanth improves the handling and sheeting properties of icing and is invaluable in sugarcraft, the preparation of cake decorations from sugar.

Limitations

Tragacanth is relatively expensive and its use in dressings has been largely replaced by xanthan gum. Dispersions of tragacanth can take a long time to hydrate fully unless high shear mixing is used. Measures may need to be taken to avoid lumping when adding to water. Within the EU, tragacanth is permitted in Group 1, Part C of Regulation 1129/2011, additives permitted at *quantum satis*. It is not permitted in jelly minicups.

Typical Products
Confectionery icing, pourable and spoonable dressings and flavour oil emulsions.

Number	Product
E 414	Acacia gum
	Gum arabic

Sources
Gum arabic is a gummy exudate produced by trees of the species *Acacia senegal* (L.) Willd. and its close relatives as a response to wounding. The majority of the trees are wild but there are some orchards, mainly in the Sudan. Gum production is encouraged by making a transverse incision in the bark of the trunk and peeling off a thin strip of bark. The gum appears as pale yellow orange tears about the size of a table tennis ball that harden rapidly by evaporation. The tears are collected by hand and cleaned from loose detritus. Top-quality gum is finally cleaned by dissolving in water followed by filtration and recrystallising or spray drying to produce a powder.
The gum is a polysaccharide with a backbone of D-galactose with D-glucuronic acid units and L-rhamnose or L-arabinose end units.

Function in Food
Gum arabic is used as a viscosity modifier and emulsion stabiliser.

Benefits
Gum arabic is very soluble in water (solutions of up to 50% can be obtained) with a pH of 4.5–5.5. It is practically colourless, odourless and tasteless and imparts mouthfeel without gumminess. It is effective in keeping oils in suspension without large increases in viscosity and particularly for encapsulating flavouring oils both for soft drinks and for spray drying to produce powdered flavours. In soft drinks it allows a long shelflife and for the dried product it gives good content to shell ratios and a clean flavour.
Gum arabic can also be regarded as a source of soluble fibre, being unaffected by passage through the stomach but broken down by the large intestine.
It is also used to inhibit sugar crystallisation in sweets.
Gum arabic has a property of forming coacervates with gelatin that forms the basis of its use as a wall material for microencapsulation.

Limitations
Acacia gum is permitted in the EU in Part C Group 1 of Regulation 1129/2011, additives permitted at *quantum satis*, but it is not permitted in jelly minicups.

Being a natural product, gum arabic supply is liable to considerable fluctuation and it is increasingly being replaced by modified starches.

Gum arabic is less effective at generating viscosity than most other gums and thickeners.

Typical Products
Soft drinks, confectionery gums and spray-dried flavours.

Number	Product
E 415	Xanthan gum

Sources
Xanthan gum is a polysaccharide produced by the fermentation of sugars by the bacterium *Xanthamonas campestris*, which was originally found growing on cabbage leaves. At the end of the fermentation, the broth is sterilised and the gum isolated by precipitation with propanol before washing and drying.

Function in Food
Xanthan gum is used to increase viscosity in sauces and dressing, drinks and cakes. It is particularly stable to acid, heat and enzymes, resulting in no loss of viscosity over the shelflife of the products.

Benefits
Solutions of xanthan gum are thick/viscous when at rest but become thinner when they are stirred. The viscosity is regained immediately the stirring stops. This means that they can be used to hold particles in suspension, but the solution will flow easily on stirring or pumping.

In sauces and dressings, xanthan gum is used to provide body and mouthfeel, to increase stability to acid and heat, to provide tolerance to repeated freezing and thawing, and to aid emulsion stability. A useful property is that, when sauces containing xanthan gum are poured out of a bottle, they cut off cleanly and do not drip.

In drinks, xanthan is used to improve mouthfeel, particularly in diet products, and to hold particles such as cocoa and orange pulp in suspension. It forms synergistic mixtures with guar and locust bean gums.

A major use is in baking where xanthan is used to reduce splashing during filling moulds, to hold particles such as chocolate chips in suspension where the batter is fluid, and to increase volume in the finished product.

It is also finding increasing use in gluten-free bakery products such as gluten-free bread where it has been shown to improve the texture and shelflife of products.

Limitations
In the EU, xanthan gum is included in Part C Group 1 of Regulation 1129/2011, additives permitted at *quantum satis*, but it is not permitted in jelly

minicups. Because it readily absorbs water and swells, it should not be ingested as a dry powder and is not permitted in dehydrated foods intended to rehydrate on ingestion. Its rapid rate of hydration means that it is important to ensure that it is well dispersed throughout a mix before water is added, or it can form lumps.

Typical Products
Sauces and dressings, drinks, cakes, fruit preparations, desserts, meat products and gluten-free products.

Number	Product
E 416	Karaya gum

Sources
Karaya gum is a natural exudate collected from trees of *Sterculia urens* (Roxburgh) and other species of *Sterculia* and *Cochlospermum* which grow in India, Senegal and Mali. Incisions are made in the bark of the tree and the gum is exuded from the cut. The tears are allowed to dry before being collected, cleaned from foreign matter, sorted by colour and milled to a powder. Different grades are classified by colour, particle size and viscosity. Karaya gum is a high molecular weight (5 to 8 million Daltons), branched anionic polysaccharide, which occurs as a partially acetylated, mixed calcium and magnesium salt. The structure of karaya gum is not fully understood, but appears to consist of a backbone based on D-galacturonic acid and L-rhamnose with side chains of D-galactose and D-glucuronic acid. The ratio of these constituents varies depending on the source of karaya. Overall, karaya contains approximately 37% uronic acid residues and 8% acetyl groups.

Function in Food
Karaya gum is used as a thickener and as a coating and glazing agent. Karaya particles do not normally dissolve but swell in a similar fashion to starch, although karaya generally thickens at a lower concentration than starch, forming a thick paste at 3% w/w in water.

Benefits
Karaya is useful as a thickener since it does not have the "gummy" texture associated with many other hydrocolloids. It can also provide better flavour release than the equivalent level of starch. The texture of a karaya paste in water can, to some extent, be controlled by the original particle size of the dry powder. Karaya gum has good acid stability and is resistant to hydrolysis by food enzymes. Its indigestibility has resulted in its use as a laxative.

Limitations
The acidic flavour of karaya gum has limited the number of applications for the gum. In order to achieve maximum viscosity, karaya gum should be dispersed in water prior to the addition of other ingredients such as acid or sugar. Karaya

gum is not normally used at a pH higher than 7 since its rheology changes to a ropey mucilage as a result of deacetylation in alkaline conditions. The stability of karaya gum in powder form (with respect to water viscosity) is not as good as that of some other hydrocolloids. Within the EU, karaya gum is included in Regulation 1129/2011, where it is permitted in a range of foodstuffs with individual maxima in each case.

Typical Products
Sauces, in particular brown sauce, coatings, fillings, toppings and chewing gum.

Number	Product
E 417	Tara gum

Sources
Tara gum is the ground endosperm from the seed of the tara shrub *Caesalpinia spinosa*, which is indigenous to Peru. The main component of the powder (*ca.* 80%) is a high molecular weight linear polysaccharide (galactomannan) with a backbone chain of 1-4 linked β-D mannose residues, to which 1-6 α-D-galactose sugars are attached. The distribution of the galactose along the mannan chain is not known, but statistically there is approximately one galactose residue for every three mannose residues.

Function in Food
Tara gum is an efficient thickening and gelling agent. It dissolves partially in cold water, generating *ca.* 70% of its potential functionality. It hydrates fully in water above 85 °C, forming an opalescent pseudoplastic solution. Mixed with xanthan, it forms thermoreversible gels and increases the elasticity of kappa-carageenan gels.

Limitations
Within the EU, tara gum is included in Part C Group 1, additives permitted at *quantum satis*, in Regulation 1129/2011, but it may not be used for jelly minicups. Because it readily absorbs water and swells, it should not be ingested as a dry powder and it is not permitted for dehydrated foods intended to rehydrate on ingestion.

Typical Products
Sauces, soups, ketchups and mayonnaises.

Number	Product
E 418	Gellan gum

Sources
Gellan gum is an extracellular polysaccharide prepared by fermentation of the micro-organism *Sphingomonas elodea*, previously classified as *Pseudomonas elodea*.

Gellan gum is extracted from the fermentation medium after pasteurisation, by alcohol precipitation to yield the high-acyl form or by treatment with alkali followed by alcohol precipitation to yield the unsubstituted low-acyl form. The gum is a mixed salt, predominantly the potassium salt but also containing other cations such as sodium, calcium and magnesium.

Function in Food
Gellan gum is soluble in hot water and is used for gelling, stabilising or film forming, with the precise properties depending on the degree of acyl substitution. The gels form when hot solutions are cooled in the presence of gel-promoting cations such as sodium, potassium, magnesium and calcium. The gelation and hydration of the high-acyl form are less dependent on the presence of ions than is the case with the low-acyl form. Calcium, in particular, inhibits the hydration of the low-acyl form but both will hydrate in hot milk without the need for a sequestrant.
Low-acyl gum typically forms a firm, brittle gel at between 30 °C and 50 °C, while the high-acyl gum gives soft, elastic gels at around 70 °C. Intermediate textures can be produced by mixtures of the two forms of the gum. To obtain optimum gel properties it is sometimes necessary to add extra cations, usually in the form of a soluble calcium salt. This is best done when the solution is hot.
At very low concentrations, gellan gums exhibit a weak gel structure that can be sheared to produce smooth, pourable structured liquids that can be used to provide a suspension of particulates such as jelly beads, herbs or spices.

Benefits
Gellan gum is effective at very low concentrations and does not mask flavours in foods. Using the two forms of the gum in combination allows a wide range of textures to be produced. The gels are stable over a range of pH from 2.5 to 10 and can be formed over a range of concentrations up to 75%. The pourable gels allow the development of products that hold particulates in suspension but that can be consumed as drinks.

Limitations
As with all hydrocolloids, care needs to be exercised in hydrating gellan gum. Careless addition to water leads to clumping, where the outside of the powder hydrates quickly, preventing the powder inside the clump from dissolving. This difficulty can be overcome either by the use of the appropriate mixing equipment or by dry blending the gum with other ingredients, in particular sugar, prior to addition to water.
Low acyl gels are not thermally reversible and all gellan gels are susceptible to degradation in hot acidic conditions so that addition of acid needs to be as late in a process as possible.
Within the EU, gellan gum is included in Group 1 of Part C of Regulation 1129/2011, additives permitted at *quantum satis*. In common with other gelling agents it is not permitted in jelly minicups.

Typical Products
Fruit fillings and bakery jams, jelly drinks.

Number	Product
E 420	Sorbitol
	Crystalline sorbitol
	Sorbitol syrup

Sources
Sorbitol is widely present in nature, particularly as a constituent of many fruits and berries. Commercial products are manufactured by the hydrogenation of dextrose syrup, followed, for the crystalline sorbitol, by crystallisation.

Function in Food
Sorbitol is available as a pure crystalline material and as aqueous solutions having a dry matter content of 70%. Sorbitol is a nutritive sweetener and replaces sucrose and glucose syrups, for bulk, texture and sweetness, in sugar-free confectionery products such as chewing gum, compressed tablets and hard-boiled, soft and chewy candies. Sorbitol syrup is also used as an efficient humectant, and as a sequestering and emulsifying agent in confectionery and bakery products, as well as in mayonnaise, creams and sauces. It is also used to inhibit crystallisation of other sugars.

Benefits
Sorbitol does not promote tooth decay and has a reduced calorie content (2.4 kcal/g in Europe, 2.6 kcal/g in the USA). It extends the shelflife of food products and does not provide browning in food when heated or baked. Sorbitol can be combined with other polyols as well as with intense sweeteners to balance its slightly reduced sweetening power (ca. 60% that of sucrose). Sorbitol is well tolerated by diabetics.

Limitations
In the EU sorbitol is permitted in Part C Group 4 (polyols) of Regulation 1129/2011. It is permitted at *quantum satis* in table-top sweeteners and in a range of energy-reduced or no-added-sugar products, for example confectionery, dietary products and supplements. It is also permitted for purposes other than sweetening in fresh fish and crustacea.
It has been allocated an ADI of "not specified" by JEFCA.
As with other polyols, if a foodstuff contains more than 10% sorbitol, the product must be labelled to the effect that excessive consumption may produce laxative effects.

Typical Products
Chewing gum, sugar-free confectionery, bakery products and surimi.

Number	Product
E 421	Mannitol

Sources
Mannitol is widely present in nature, particularly in fruits, plants and algae. Commercial mannitol is manufactured by hydrogenation of fructose or mannose, followed by crystallisation and drying.

Function in Food
Mannitol is a nutritive sweetener with 50–60% of the sweetness of sugar. It is often used in combination with other polyols or other intense sweeteners.
It is used as a polyol and to control water activity in order to reduce stickiness in chewing gum and hard boiled candies.

Benefits
Mannitol is not metabolised by the bacteria that cause tooth decay and is not metabolised by the human body in the same way as sugar. It has been given a calorific value of 1.6 kcal/g in the USA and 2.4 kcal/g in the EU. It extends the shelflife of food products and does not participate in the Maillard reaction.

Limitations
Mannitol is one of the polyols permitted in the EU in Group IV Part C of Regulation 1129/2011. It is permitted at *quantum satis* in table-top sweeteners and in a range of energy-reduced or no-added-sugar products, for example confectionery, dietary products and supplements. It is also permitted for purposes other than sweetening in fresh fish and crustacea.
As with other polyols, if a foodstuff contains more than 10% mannitol, the product must be labelled to the effect that excessive consumption may produce laxative effects.

Typical Products
Sugar-free chewing gum, sugar-free hard-boiled candies and chocolate.

Number	Product
E 422	Glycerol

Sources
Glycerol is made by the hydrolysis of fats. It can be obtained from both animal and vegetable fats, and material from both sources is readily available.

Function in Food
Glycerol is a clear, almost colourless, liquid at room temperature. It is used as an humectant, to keep foodstuffs moist to the palate without the risk of mould or bacterial growth. It is also used to retard staling and to improve texture by plasticising the food.

Benefits
Glycerol is naturally present in food and is formed in the human digestive system. It is readily available and has a long history of use.
Because it is liquid at room temperature it is used to replace water, keeping product moist to the palate while reducing the water activity. In products where sugar crystallises after manufacture, glycerol is used to inhibit crystallisation, thus maintaining more sugar in solution, which itself has an humectant effect. Glycerol is also less volatile than water, which means that it is better at maintaining moistness over the shelflife of the product.

Limitations
Glycerol has a particular taste effect of leaving a slight burning sensation in the throat that limits the amount that can be used in a product. Glycerol is permitted in the EU in Part C Group 1 of Regulation 1129/2011, additives permitted at *quantum satis*.
It is listed as GRAS in the USA.
It has been allocated an ADI by JECFA of "not specified".

Typical Products
Cakes and confectionery.

Number	Product
E 425	(i) Konjac gum
	(ii) Konjac glucomannan

Sources
Konjac gum and konjac glucomannan, also known as konjac flour, yam flour, konnyaku glucomannan and glucomannan gum, are extracted from the tubers of the plant *Amorphophallus konjac* plant. Tubers are harvested after 2–3 years, when they contain 30–50% glucomannan, which is sufficient for commercial extraction. After harvesting, the tubers are washed and cleaned quickly to avoid bruising and spoilage, followed by slicing and chipping to assist drying. The dried tubers are ground and separated by air classification. The heavier idioblast sacs, which contain the konjac gum, are recovered and washed with alcohol and water to remove starch, protein and other unwanted materials together with the strong fishy taints naturally associated with konjac. Finally the powder is dried, ground and blended.

Function in Food
Konnyaku noodles are a traditional food in the Far East, made by heating glucomannan solutions with limewater (E 526) to form a thermally stable gel, which is cut into thin strips and used as a meal component. In table dessert gels and aspics, 0.6% of a konjac gum–kappa carrageenan blend gives firm cohesive textures. The thermally stable glucomannan gel is used in coarse ground sausage and meat analogues as a texture modifier and water binder. Konjac

gum acts as a binder and protects against freezer damage in surimi. In cream cheese and processed cheese, a low level of around 0.2% glucomannan is very effective for moisture binding and good spreading properties and for giving a creamy mouthfeel and full body. Konjac gum provides ice crystal control, thickening and bodying to ice cream and frozen desserts. It is used in sauces, gravies, salad dressings and mayonnaise for thickening and stabilising. In bakery applications, the glucomannan acts as a film former and flow aid for coatings, toppings and batter and as a binder and extrusion aid for pasta.

Benefits
The high molecular weight of around 1 million Daltons for this linear glucomannan confers a high viscosity when the gum is fully hydrated. The nonionic D-mannose and D-glucose units in konjac gum are relatively unaffected by high levels of salt, and the glucomannan is stable to below pH3.8. The gum is a source of soluble fibre as the β1-4 linkages in the glucomannan chain resist enzymic degradation during digestion.
The glucomannan contains random acetyl groups, which prevent long-chain polymers from associating to form a gel. The acetyl groups can be removed by adding a weak base to raise the konjac solution above pH9, and heating. Once the side groups are removed, the polymer chains interact to form nonmelting gels. The rate of gel formation is controlled by pH and temperature. Gelation proceeds as the gel is deacetylated so that a gel may be formed at any temperature: there is no specific setting temperature as in the case of carrageenan, agar or gelatine. Gels are insoluble in water and are stable to temperatures above 200 °C.
Adding 0.02 to 0.03% konjac flour to 1.0% xanthan gum will increase the viscosity two to three times through interchain associations between the two polymers. Higher levels of konjac will form a thermally reversible gel with xanthan. Blends of konjac and kappa-carrageenan show stronger synergy than blends of carrageenan and locust bean gum. Heat is required to hydrate both gums fully and a thermally reversible gel forms upon cooling. By varying the gum ratio, the texture can be varied.

Limitations
Konjac has been consumed in foods for over 1000 years in Asia, and is considered a food product. To the rest of the world, it is a relatively new food ingredient. In the EU, in Regulation 1129/2011 it is included in Part C Group 1, only being limited to 10 g/kg individually or in combination. It is not permitted in jelly minicups nor in jelly confectionery. Konjac gum hydrates in water at room temperature but heating or shearing the solution greatly speeds up this process.

Typical Products
Aspics, frozen desserts, sauces and surimi.

Number	Product
E 426	Soyabean hemicellulose

Sources
Soyabean hemicellulose is a free-flowing white to yellowish powder obtained by hot-water extraction from soya fibre, which is in turn a byproduct of the process for the production of soya oil and soya protein. A number of products are available that differ in viscosity.

Function in Food
Soyabean hemicellulose is soluble in both cold and hot water. It has a number of uses. It has good water-binding capacity and is used as a thickener in baked goods intended for freezing and warming by microwave and in jelly confectionery. It stabilises protein in acidic conditions and is used as a stabiliser in yogurt drinks. In nonfried instant noodles, it improves the texture of the dough and shortens the cooking time of the noodles. It is also used to control the stickiness of the surface of cooked rice or noodles preventing them from becoming too glutinous and keeping their glossy appearance, which improves their mixing characteristics with other added ingredients.

Limitations
Within the EU, soyabean hemicellulose is permitted in Regulation 1129/2011 in a number of products with individual maxima in each case; prepacked processed potato products, jelly confectionery except minicups, prepacked ready to eat rice, frozen egg products, emulsified sauces and food supplements.

Typical Products
Dairy based drinks, oriental noodles and rice products.

Number	Product
E 427	Cassia gum

Sources
Cassia gum is the ground purified endosperm of *Cassia obtusifolia* or *Cassia tora*, bushy shrubs that grow wild in the India subcontinent. Seeds of *Cassia occidentalis* are natural contaminants and are reduced to no more than 0.05% during the initial cleaning process. The clean product is dehusked and de-germed, milled and ground to produce the material of commerce.
The gum is a polysaccharide with a molecular weight of 200 000 to 300 000 Daltons and composed mainly of galactomannans with a mannose:galactose ratio of approximately 5:1.

Function in Food
Cassia gum is a thickener and gelling agent. It dissolves partially in cold water but attains its full viscosity in hot water.

It has a lower ratio of galactose to mannose than other seed gums that makes it more effective in combination with carageenan or agar than the others. It forms firm and thermoplastic gels with carageenan and agar, and soft viscoelastic gels with xanthan.

Benefits
Cassia has been used in pet food for many years but has only recently (2010) been permitted for use in human foods so the current uses are limited. However, it is used in dairy products and heat-treated meat products.

Limitations
Cassia gum has a more limited range of permitted uses than the other seed gums but is permitted in EU Regulation 1129/2011 for uses in flavoured milks, edible ices, fillings, toppings and coatings for fine bakery wares and desserts, heat-treated processed meat, dehydrated soups and broths, sauces and dairy-based desserts with individual limits.

Typical Products
Flavoured milk and dairy desserts.

Number	Product
E 431	Polyoxyethylene (40) stearate

Sources
Polyoxyethylene (40) stearate is made by reacting stearic acid with polyoxy-ethylene, a polymer of ethylene oxide.

Limitations
Within the EU, in Regulation 1873/84 it is only permitted in some imported wines at levels not exceeding 18 mg/l. It is not permitted in wine made within the EU.

Number	Product
E 432	Polyoxyethylene sorbitan monolaurate
	Polysorbate 20

Sources
Polysorbate 20 is a pale-yellow liquid produced from a mixture of partial laurate esters of sorbitol and its anhydrides, condensed with ethylene oxide.

Function in Food
Polysorbate 20 is widely used within the food industry as a surfactant, for forming oil-in-water emulsions such as dressings, sauces and margarines. The surfactant properties also lead to uses in improving the volume and texture of

cakes, the dispersion of coffee whiteners, and the aeration, dryness and texture of whipped cream.

Benefits
Polysorbate 20 can be used in combination with mono- and diglycerides of fatty acids (E471) or other soluble polysorbates to provide the optimum balance of emulsion properties. It is soluble in hot and cold water but insoluble in edible oils.

Limitations
Polysorbate 20 is one of the polysorbates included in Regulation 1129/2011 in the EU where they are permitted in a range of product categories, with limits in each case.
Polysorbate 20 has a warm, somewhat bitter taste.

Typical Products
Cakes and cake mixes, dairy based whipped cream, salad dressing and coffee whiteners.

Number	Product
E 433	Polyoxyethylene sorbitan monooleate
	Polysorbate 80

Sources
Polysorbate 20 is a pale-yellow liquid produced from a mixture of partial oleate esters of sorbitol and its anhydrides, condensed with ethylene oxide.

Function in Food
Polysorbate 80 is used as a surfactant, often in combination with other emulsifiers for forming oil-in-water emulsions. It is used to stabilise margarine, sauces and dressings, and to hold the fat in ice cream.

Benefits
Polysorbate 80 can be used in combination with mono- and diglycerides of fatty acids (E471) or other soluble polysorbates to provide the optimum balance of emulsion properties. It is soluble in hot and cold water but insoluble in edible oils.

Limitations
Within the EU, polysorbate 80 is one of the polysorbates included in Regulation 1129/2011 where they are permitted in a range of product categories, with individual maxima in each case.

Typical Products
Ice cream, frozen desserts, margarine.

Number	Product
E 434	Polyoxyethylene sorbitan monopalmitate
	Polysorbate 40

Sources
Polysorbate 40 is a pale-yellow liquid produced from a mixture of partial palmitate esters of sorbitol and its anhydrides, condensed with ethylene oxide.

Function in Food
Polysorbate 40 is a surfactant, often used in combination with other emulsifiers for forming oil-in-water emulsions. It is used to stabilise sauces and dressings, and in bakery margarine to improve aeration, cake volume and texture. It is also used in coffee whiteners and in whipped cream.

Benefits
Polysorbate 40 can be blended with other polysorbates or with E 471 to provide optimum balance of emulsion properties for particular uses. It is soluble in hot and cold water but insoluble in edible oils.

Limitations
Poysorbate 40 is permitted in the EU as one of the polysorbates in Regulation 1129/2011 where they are permitted in a range of products with limits in each case. Like other polysorbates, polysorbate 40 has a warm, somewhat bitter taste.

Typical Products
Cakes and cake mixes, margarine, coffee whiteners, sauces.

Number	Product
E 435	Polyoxyethylene sorbitan monostearate
	Polysorbate 60

Sources
Polysorbate 60 is a pale-yellow liquid produced from a mixture of partial stearate esters of sorbitol and its anhydrides, condensed with ethylene oxide.

Function in Food
Polysorbate 60 is a surfactant, often used in combination with other emulsifiers for forming oil-in-water emulsions. It is used in bakery margarine to improve dough conditioning and reduce staling in bread and to improve batter aeration in cakes. It is also used for dressings and sauces.

Benefits
Polysorbate 60 can be blended with other polysorbates or with E 471 to provide the optimum balance of emulsion properties. It is soluble in hot and cold water but insoluble in edible oils.

Limitations
Polysorbate 60 is one of the polysorbates permitted in the EU in Regulation 1129/2011 where they are permitted in a range of product categories with limits in each case. Like other polysorbates, polysorbate 60 has a warm, somewhat bitter taste.

Typical Products
Cakes and cake mixes, coffee whiteners, margarine and salad dressings

Number	Product
E 436	Polyoxyethylene sorbitan tristearate
	Polysorbate 65

Sources
Polysorbate 65 is a tan colour solid produced from a mixture of partial stearate esters of sorbitol and its anhydrides, condensed with ethylene oxide.

Function in Food
Polysorbate 65 is used as a surfactant, often in combination with other emulsifiers for forming oil-in-water emulsions. It is used to hold the fat in ice cream to give dry eating characteristics and to retard the development of fat bloom in chocolate products. It is used to retard foam formation during food processing.

Benefits
Polysorbate 65 can be used in combination with E471 or other polysorbates to provide the optimum balance of emulsion properties. It is soluble in hot and cold water but insoluble in edible oils.

Limitations
Polysorbate 65 is permitted in the EU as one of the group of polysorbates included in Regulation 1129/2011 where they are permitted in a range of product categories, with individual limits in each case.
Polysorbate 65 has a waxy, somewhat bitter taste.

Typical Products
Ice cream and frozen desserts, whipped creams and cakes.

Number	Product
E 440	Pectins
	(i) pectin
	(ii) amidated pectin

Sources
Pectins are found in most land plants, especially in fruits and other nonwoody tissue. Commercial pectins are currently extracted from fruit solids remaining

after juice extraction – in particular, from apple pomace and citrus peel. Other minor sources are sugar-beet pulp after the removal of sugar, and sunflower-head tissue after removal of the seeds. The choice of source material is determined by availability on a sufficient scale, and by the suitability of the pectins obtained for use in food additive and ingredient functions.

Function in Food
Pectins are used as gelling and thickening agents in a range of mainly acidic foods, most typically fruit products (jams, jellies, industrial fruit preparations for bakery and dairy products, sugar confectionery) but also increasingly in glazes and sauces for savoury products. High methoxyl pectins are also used as stabilisers of proteins in acidic products such as yogurts and soya analogues, where heat treatment is required, in ice pops and sorbets, and to improve mouthfeel in drinks (especially low calorie or low fruit). Low methoxyl pectins are also used to gel or thicken desserts, either water or milk based.

Benefits
Pectin derived from fruit is the obvious gelling agent to supplement the natural pectin in fruit products. In confectionery, it gives a clear tender gel with good flavour release, which requires no stoving process after depositing. In low-sugar fruit bases, amidated low-methoxyl pectin can give a thixotropic texture, which is pumpable but capable of suspending fruit pieces. Amidated pectin produces completely thermally reversible gels, whilst nonamidated low-methoxyl and high-methoxyl pectins give gels with considerable resistance to melting, and hence bakefast properties. Pectin is an effective stabiliser for acidic protein systems, which does not give excessive viscosity, and is therefore ideal for yogurt and similar drinks.
Pectins are one form of soluble dietary fibre, and may be used to increase the fibre content of suitable foods and drinks.

Limitations
Both pectin and amidated pectin are considered GRAS in the USA and are permitted in the EU in Part C, Group 1 of Regulation 1129/2011, additives permitted at *quantum satis*, except that they are not permitted in jelly minicups. They have been allocated an ADI of "not specified" by JEFCA.

Typical Products
Jam, marmalade, sugar confectionery, fruit bases and soft drinks.

Number	Product
E 442	Ammonium phosphatide
	Emulsifier YN

Sources
Ammonium phosphatide is obtained by phosphorylation of a mono- and diglyceride produced from an edible fat. Traditionally, a partially

hydrogenated rapeseed oil is used as the fat source. After the phosphorylation with phosphorus pentoxide, the product is neutralised with ammonia, forming a mixture of ammonium salts of phosphatidic acids.

Function in Food

Ammonium phosphatide is an emulsifier mainly used for adding to chocolate in order to reduce the viscosity of the liquid chocolate, thus making it suitable for further processing such as moulding or enrobing. In chocolate, ammonium phosphatide is found on the surface of the particles, especially on sugar particles, so the friction between the particles is reduced. Ammonium phosphatide also works by dispersing agglomerated particles during the conching process.

Ammonium phosphatide is also added to couverture, ice cream coatings and various confectionery products, where it can be used as a substitute for lecithin.

Benefits

Ammonium phosphatide has a neutral flavour profile and does not add any off-flavours to the food products, even when added at high dosages up to 1%. Ammonium phosphatide provides a higher stability against oxidation than, for example, lecithin.

Its ability to control viscosity makes it possible to reduce the fat content of the final products. In chocolate, ammonium phosphatide works synergistically with the emulsifier PGPR (E 476), enabling the manufacturer to obtain an additional saving in the amount of cocoa butter added.

Limitations

Ammonium phosphatide has been allocated an ADI of 0–30 mg/kg body weight by JEFCA and is considered GRAS in the USA. In the EU, it is permitted in Regulation 1129/2011 at levels of up to 10 000 mg/kg in chocolate- and cocoa-based confectionery.

Typical Products

Chocolate, couverture.

Number	Product
E 444	Sucrose acetate isobutyrate (SAIB)

Sources

Sucrose acetate isobutyrate is produced by the controlled esterification of sucrose using acetic and isobutyric acid anhydrides. The precise pattern of esterification will depend on the reaction conditions. The molecular weight can vary between 832 and 856. It is a very viscous, clear, colourless liquid.

Function in Food

In the preparation of cloudy, flavoured soft drinks, essential oils are often used in an emulsion as part of the flavouring system. SAIB is used to inhibit the coalescence and separation of the oils from the body of the drink. The oils generally have a lower density than the water in the drink and can, unless some preventive action is taken, separate out at the top of the container. SAIB both increases the density of the oil and acts to stabilise the emulsion, usually in conjunction with other water-phase additives, such as gum arabic (E 414). The stabilisation is also believed to be aided by charges on the emulsion droplets generated during the emulsification process.

Benefits
SAIB is flavourless and odourless at the levels used in beverages, and is stable to oxidation. It metabolises to sugar, acetic and isobutyric acids.

Limitations
At room temperature SAIB is a very viscous liquid and must either be warmed to 60 °C or mixed with orange terpenes before use. In the EU, SAIB is permitted in Regulation 1129/2011 only for use in nonalcoholic flavoured cloudy drinks and flavoured cloudy drinks containing less than 15% alcohol by volume, up to a maximum of 300 mg/l.

Typical Products
Cloudy soft drinks

Number	Product
E 445	Glyceryl esters of wood rosin
	Ester gum

Sources
Wood rosin is a pale yellow, acidic material extracted from pine wood chips. The major component is abietic acid. The rosin is reacted with glycerol (E422) to give a mixture of di- and triglycerides, which is purified by countercurrent steam distillation to yield a hard, clear, pale yellow thermoplastic resin.

Function in Food
In the preparation of cloudy, flavoured soft drinks, essential oils are often used in an emulsion as part of the flavouring system. Ester gum is used to modify the properties of the oils so that they remain evenly distributed throughout the drinks during their shelflife. It is believed to act by increasing the density of the oil, acting as a stabiliser and through charges generated on the droplet surface during the emulsification process.

Benefits
Ester gum is odourless and tasteless at the concentrations used. It is available as small beads, which allows for improved dispersion when preparing a solution.

Limitations
Within the EU, ester gum is only permitted in Regulation 1129/2011 to a maximum level of 100 mg/l in three categories of cloudy drinks; nonalcoholic,

spirit-based or containing less than 15% alcohol and as a surface treatment for citrus fruit at no more than 50 mg/kg.

Typical Products
Cloudy soft drinks

Number	Product
E 450	Diphosphates
	(i) disodium diphosphate
	(ii) trisodium diphosphate
	(iii) tetrasodium diphosphate
	(iv) tetrapotassium diphosphate
	(v) dicalcium diphosphate
	(vi) calcium dihydrogen diphosphate

Sources
The original source of the diphosphates is phosphate rock, which is mined in areas such as Morocco, Israel, North America and Russia. Yellow phosphorus is extracted from phosphate rock using either a high-energy electrothermal process or an acid extraction. The phosphorus is burnt in an oxygen atmosphere at very high temperatures to produce phosphorus pentoxide. This is dissolved in dilute phosphoric acid and reacted with sodium, potassium or calcium hydroxide to produce an orthophosphate. The orthophosphates are then combined in a high-temperature condensation reaction to form chains of two phosphate units – the diphosphates.
Sodium, potassium and calcium diphosphates are available. The diphosphates used in food applications are the disodium, trisodium, tetrasodium, tetrapotassium, dicalcium and calcium dihydrogen forms.

Function in Food
The baking industry is the largest user of the diphosphates, where their principal function is that of leavening agent. The acidic diphosphates are used in this application, the most widely used being sodium acid pyrophosphate (disodium diphosphate), which is usually known by its initials SAPP, and the calcium diphosphates.
The phosphates function as stabilisers in meat products, where they work synergistically with salt, interacting with the meat fibres and causing fibres to expand and retain water within them. They also work with salt to extract the meat proteins, allowing the formation of a meat protein exudate, which will bind meat pieces together in a comminuted or reformed product.
In processed cheese, cheese preparations and cheese-based sauces, the phosphates act as emulsifying salts. In this application, they break the calcium bridges between the cheese protein molecules by means of ion exchange, converting the insoluble cheese protein complexes into individual soluble protein molecules. These protein molecules are then able to emulsify the fat associated with the cheese, in a manner similar to that of sodium caseinate.

As this interaction relies on the exchange of sodium or potassium for the calcium associated with the cheese proteins, the calcium phosphates cannot function in this way.

The diphosphates can aid gel formation in products such as instant whips.

Benefits

At least five grades of SAPP are commercially available, differing in their rate of reaction with sodium bicarbonate for use as raising agents. The slower grades are used in large cakes and refrigerated doughs, where consistency of gas release over a long period of time is required, while the faster grades are used in cake doughnuts and small cakes. The use of SAPP increases the alkalinity of the finished cake and increases the rate of browning compared with the use of monocalcium phosphate.

In meat products, the phosphate and salt interaction extracts salt-soluble protein, which binds individual meat pieces together. The meat fibres also expand, allowing greater retention of meat juices, thereby improving succulence.

Without the use of emulsifying salts, such as the phosphates, it is impossible to produce stable processed cheese or cheese-based sauces.

Limitations

The diphosphates are fast acting in meat, but they are the least soluble and application is more difficult. They require vigorous action in order to incorporate them into meat systems. They are less suitable for mince-mix systems, where a blend of diphosphates with triphosphates and/or polyphosphates is recommended.

The use of SAPP as a raising agent can result in a distinct aftertaste, which can be minimised by careful adjustment of the acid to bicarbonate ratio.

Calcium phosphates have poor solubility and this limits their application in many food types.

In the EU, the diphosphates are included, in Regulation 1129/2011, in the group of phosphates that, individually or in combination, are permitted in a wide range of products with individual limits in each case.

Typical Products

Meat products, processed cheese, bakery products, beverage whiteners, UHT and sterilised milk products.

Number	Product
E 451	Triphosphates
	(i) pentasodium triphosphate
	(ii) pentapotassium triphosphate

Sources

The original source of the triphosphates is phosphate rock, which is mined in areas such as Morocco, Israel, North America and Russia. Yellow phosphorus is extracted from phosphate rock using either a high-energy electrothermal process or an acid extraction. The phosphorus is burnt in an oxygen atmosphere at very high temperatures to produce phosphorus pentoxide. Phosphorus pentoxide is dissolved in dilute phosphoric acid and reacted with sodium, potassium or calcium hydroxide to produce an orthophosphate. The orthophosphates are then combined in a high-temperature condensation reaction to form chains of three phosphate units – the triphosphates.

Two triphosphates are available – pentasodium triphosphate (sodium tripolyphosphate) and pentapotassium triphosphate. The pentasodium form is widely available and widely used; the pentapotassium form is less common.

Function in Food
The phosphates function as stabilisers in meat products, where they work synergistically with salt, interacting with the meat fibres and causing fibres to expand and retain water within them. They also work with salt to extract the meat proteins, allowing the formation of a meat protein exudate, which will bind meat pieces together in a comminuted or reformed product.

In fish and seafood processing, the triphosphates (and polyphosphates) substantially reduce the drip loss on storage, maintaining the succulence of the products and avoiding the dry and fibrous texture otherwise encountered. In contrast to their functionality in meat products, in this application, the phosphates work both with and without salt.

In processed cheese, cheese preparations and cheese-based sauces, the phosphates act as emulsifying salts. In this application, they break the calcium bridges between the cheese protein molecules by means of ion exchange, converting the insoluble cheese protein complexes into individual soluble protein molecules. These protein molecules are then able to emulsify the fat associated with the cheese, in a manner similar to that of sodium caseinate.

Benefits
In meat products, the phosphate and salt interaction extracts salt-soluble protein, which binds individual meat pieces together. The meat fibres also expand, allowing greater retention of meat juices, thereby improving succulence.

The triphosphates do not act as quickly as the diphosphates on the meat proteins; they must first be broken down to the diphosphate form by enzymes in the meat. They are, however, more soluble and better suited to dissolution in brine for injection or tumbling of meat, and are far more suitable for mince-mix systems. The optimum results are achieved when they are used as a blend with diphosphates and/or polyphosphates.

Without the use of emulsifying salts, such as the phosphates, it is impossible to produce stable processed cheese or cheese-based sauces.

As with most phosphates types, the triphosphates are often combined with diphosphates and/or polyphosphates, to give the advantages of solubility and functionality.

Limitations
In the EU, the triphosphates are included in Regulation 1129/2011, in the group of phosphates that, individually or in combination, are permitted in a wide range of products with individual limits in each case.

Typical Products
Meat products, fish and seafood, processed cheese, cheese-based sauces.

Number	Product
E 452	Polyphosphates
	(i) sodium polyphosphate
	(ii) potassium polyphosphate
	(iii) sodium calcium polyphosphate
	(vi) calcium polyphosphate

Sources
The original source of the polyphosphates is phosphate rock, which is mined in areas such as Morocco, Israel, North America and Russia. Yellow phosphorus is extracted from phosphate rock using either a high-energy electrothermal process or an acid extraction. The phosphorus is burnt in an oxygen atmosphere at very high temperatures to produce phosphorus pentoxide. This, in turn, is dissolved in dilute phosphoric acid and reacted with sodium, potassium or calcium hydroxide to produce an orthophosphate. The ortho-phosphates are then combined in a high-temperature condensation reaction to form chains of two or three phosphate units.

Further polymerisation, to produce longer chain lengths, is achieved by heating in a furnace to form a "glassy" phosphate. This is ground to give a powder, which is composed of a mixture of different chain lengths varying from four units up to thirty or more units. By varying the polymerisation conditions, it is possible to alter the average chain length. These mixtures of phosphates are grouped together under the heading of polyphosphates. Sodium, potassium and calcium polyphosphates are available. The sodium polyphosphates are widely available and widely used; the potassium and calcium polyphosphates are less common.

Function in Food
The phosphates function as stabilisers in meat products, where they work synergistically with salt, interacting with the meat fibres and causing fibres to expand and retain water within them. They also work with salt to extract the meat proteins, allowing the formation of a meat protein exudate, which will bind meat pieces together in a comminuted or reformed product.

In fish and seafood processing, the polyphosphates substantially reduce the drip loss on storage, maintaining the succulence of the products and avoiding the dry and fibrous texture otherwise encountered. In contrast to their func-tionality in meat products, in this application, the phosphates work both with and without salt.

In processed cheese, cheese preparations and cheese-based sauces, the phosphates act as emulsifying salts. In this application, they break the calcium bridges between the cheese protein molecules by means of ion exchange, converting the insoluble cheese protein complexes into individual soluble protein molecules. These protein molecules are then able to emulsify the fat associated with the cheese, in a manner similar to that of sodium caseinate. As this interaction relies on the exchange of sodium or potassium for the calcium associated with the cheese proteins, the calcium phosphates cannot function in this way. Calcium polyphosphates have poor solubility and this may restrict their function in many food types.

Benefits
In meat products, the phosphate and salt interaction extracts salt-soluble protein, which binds individual meat pieces together. The meat fibres also expand, allowing greater retention of water, thereby improving succulence. The solubility of polyphosphates increases with chain length, and polyphosphates are ideally suited to dissolution in brines for injection into meat or for use in mince-mix systems. On the other hand, they do not act as quickly as di- or tripolyphosphates. The longer chain length takes longer to convert to the diphosphate form. Optimum results in all applications are usually achieved with a blend of poly-, di- and triphosphates.
In fish and seafood processing, the phosphates help reduce drip loss and dehydration on storage, improving the succulence of the product and avoiding the dry, fibrous nature often associated with these products.
Without the use of emulsifying salts, such as phosphates, it is impossible to products stable processed cheese or cheese-based sauces.

Limitations
In the EU, the polyphosphates are included in Regulation 1129/2011 in the group of phosphates that, individually or in combination, are permitted in a wide range of products with individual limits in each case.

Typical Products
Meat products, processed cheese, fish and seafood, dried powdered foods.

Number	Product
E 459	Beta-cyclodextrin

Sources
Beta-cyclodextrin is a cyclic polymer consisting of seven D-glucose units. It is prepared by enzymic modification of starch.

Function in Food
Because of its unique "doughnut" shape, beta-cyclodextrin is able to trap other molecules and protect them against the external environment. Thus, it is used to

protect sensitive molecules against the effects of heat and light and to reduce losses through evaporation in high-temperature processes. In practice, the material to be protected is mixed with the beta-cyclodextrin in solution and then the mass is dried using a mild process such as a multistage dryer. The dry powder is then used in the product formulation. Typically the powder will contain 40% encapsulant and 60% beta-cyclodextrin.

Benefits
It is very difficult to add flavours to products made in high-temperature processes because flavour molecules tend to be volatile and are driven off during the process. Beta-cyclodextrin encapsulation can help to overcome this problem and to improve the flavour of products made by processes such as extrusion and cooking particularly by retaining more of the top notes of the flavours. It is also used to protect sensitive flavours such as orange and lime from oxidation during product storage.

Limitations
Beta-cyclodextrin is expensive and is used only where protection of flavours is important.
It is permitted in the EU in Regulation 1129/2011 in supplements in tablet form at *quantum satis*, and for use in powdered drinks up to prescribed limits. It is also permitted in Regulation 1130/2011 in encapsulated flavourings in flavoured teas, flavoured powdered instant drinks and flavoured snacks up to a maximum in each case.
In the USA it is GRAS for use as an encapsulant of flavours.

Typical Products
Sugarless confectionery, extruded snacks and frozen prepared meals.

Number	Product
E 460	Cellulose
	(i) microcrystalline cellulose
	(ii) powdered cellulose

Sources
Microcrystalline cellulose (MCC) and powdered cellulose are derived from alpha-cellulose, the most abundant natural polysaccharide found in plants and trees. Powdered cellulose is manufactured by bleaching and washing alpha-cellulose before drying and grinding to give fibres 22–110 µm in length.
Microcrystalline cellulose is manufactured by hydrolysing cellulose fibres in acid, leaving crystalline bundles. After bleaching and washing, the cellulose is dried to give agglomerates of very porous particles.
Colloidal grades of microcrystalline cellulose are formed by additional wet mechanical attrition to release individual microcrystals. To prevent reaggregation during drying so that the particles may be easily dispersed, microcrystalline cellulose is treated with a water-soluble hydrocolloid such as

carboxymethyl cellulose (E 466), guar gum (E 412), calcium alginate (E 404) or xanthan gum (E 415). The copolymer may also modify the end-use properties. Bacterial cellulose, obtained from the fermentation of *Gluconacetobacter xylinus*, formally known as *Acetobacter xylinum*, is treated with copolymers to give a range of products similar to plant-derived microcrystalline cellulose. This is used as a foodstuff in some countries but it does not have approval for food use in the EU.

Function in Food
Powdered cellulose and powdered microcrystalline cellulose are insoluble particles, which disperse readily in water. Powdered cellulose binds 4–9 times its weight of water and both materials absorb water and oil. These characteristics are the basis of their uses. Both are used to bind water to reduce stickiness and improve the extrusion properties of pasta and puffed snack foods. Powdered cellulose improves the flow properties of pancake batters and retains moisture, reduces fat uptake, and improves gas retention and crumb structure in cakes, muffins and doughnuts. The water-binding properties are used to protect against freeze–thaw damage in surimi and frozen foods. Dispersions of colloidal grades of microcrystalline cellulose are self-suspending above a critical concentration of around 0.25% and form a gel around and above 1%. Aggregates of microcrystalline cellulose and guar gum give body and creaminess to low-fat foods, such as mayonnaise, dressings and milk drinks.

Benefits
Powdered cellulose and powdered microcrystalline cellulose provide opacity and are a source of insoluble fibre in meal replacers and diet foods.
Their ability to bind water and oil allow them to be used as flavour carriers and free-flow aids in instant foods and grated cheese.
Colloidal microcrystalline cellulose is an efficient emulsion and foam stabiliser. The network is stable at all temperatures from chill to boiling point and it improves cling and coating properties. At higher concentrations, the thixotropic gel maintains the shape of extruded foods, prevents ice-crystal growth and freeze–thaw damage during storage, and holds shape when thawed. Microcrystalline cellulose has the benefit of imparting body and creaminess without gumminess.

Limitations
Powdered cellulose and microcrystalline cellulose disperse with minimal stirring, but colloidal grades must be dispersed using high shear or homo-genisation to give a stable network. The dispersion properties are not affected by temperatures. In foods with a pH value below 4.5 or more than about 1% salt (sodium chloride), microcrystalline cellulose should be dispersed in water first. A protective colloid, such as xanthan gum, at around 10% of the weight of microcrystalline cellulose, should be added to avoid flocculation and collapse of

the stabilising network. Salts or acid should be added last. Dissolved electrolytes or the presence of other water-soluble gums may extend the time for complete dispersion. In milk, it is best to disperse MCC with homogenisation above 100 bar.

In the EU, microcrystalline cellulose and powdered cellulose are permitted in Part C Group 1 of Regulation 1129/2011, additives permitted at *quantum satis*. These celluloses are included in a group that have been allocated a group ADI of "not specified" by JEFCA.

Typical Products
Puffed snack foods, baked goods, instant foods, diet foods, milk drinks, mayonnaise, dressings, frozen desserts and reformed meats.

Number	Product
E 461	Methyl cellulose

Sources
Methyl cellulose is manufactured from purified cellulose by reaction with methyl chloride under controlled conditions.

Function in Food
Methyl cellulose is soluble in cold water, where it has thickening properties, but insoluble in hot water, where gels are formed. This allows binding of food products when heated. Films of methyl cellulose exhibit good oil-barrier properties.

Benefits
The hot gelation properties of methyl cellulose are used to reduce boil-out during heating in a range of sauces and fillings. The thermal gelation properties also allow better binding and hence greatly improved shape retention in products such as reformed meats, reformed vegetables, potato products, vegetarian burgers and dietetic breads. The barrier properties can be used to reduce oil uptake in deep-fried products, both to lower fat content of the food and to reduce oil losses in processing.

Limitations
In the EU, methyl cellulose is permitted in Part C Group 1 of Regulation 1129/2011, additives permitted at *quantum satis*.
Methyl cellulose is included in the group of celluloses that have been allocated a group ADI of "not specified" by JEFCA.

Typical Products
Soya burgers, sausages, onion rings, potato croquettes, waffles and other formed potato products, gluten-free bakery products.

Number   Product
E 462   Ethyl cellulose

Sources
Ethyl cellulose is a free-flowing white to light brown powder made from wood
pulp or cotton by treatment with alkali and subsequent ethylation with ethyl
chloride, followed by purification and drying. It is insoluble in water.

Function in Food
Ethyl cellulose is used mainly in pharmaceutical preparations as a carrier for
the active substances and to provide controlled release and taste masking. It is
used in a similar role with food supplements.
It provides a good water barrier and can be used to provide a barrier between
layers in a multilayer product such as a pizza.

Limitations
Ethyl cellulose is included in the group of celluloses that have been allocated a
group ADI of "not specified" by JEFCA.
In the EU, ethyl cellulose is permitted in Part C Group 1 of Regulation
1129/2011, additives permitted at *quantum satis*.

Typical Products
Food supplements

Number   Product
E 463   Hydroxypropyl cellulose

Sources
Hydroxypropyl cellulose is manufactured by treatment of purified cellulose
with propylene oxide under controlled conditions followed by washing to
purify the product. A range of viscosities is available.

Function in Food
Hydroxypropyl cellulose (HPC) is insoluble in hot water but soluble in cold
water. Solutions vary in viscosity depending on the choice of HPC type. It is
used as a stabiliser in aerated products. It has good film-forming and barrier
properties and is soluble in ethanol.

Benefits
Hydroxypropyl cellulose can stabilise whipped toppings at high ambient
temperatures. The film-forming properties may be used to give barrier
properties, *e.g.* against oxidation. The ethanol solubility of HPC cellulose
allows it to be used for thickening of alcoholic drinks.

Limitations
Hydroxypropyl cellulose is included in the group of celluloses that have been allocated a group ADI of "not specified" by JEFCA.
In the EU, hydroxypropyl cellulose is permitted in Part C Group 1 of Regulation 1129/2011, additives permitted at *quantum satis*.

Typical Products
Aerated toppings.

Number	Product
E 464	Hydroxypropyl methyl cellulose
	Hypromellose

Sources
Hydroxypropyl methyl cellulose (HPMC) is produced from purified cellulose by treatment with methyl chloride and propylene oxide under controlled conditions, followed by washing to purify the product.

Function in Food
The properties are comparable to those of methyl cellulose, in that it is soluble in cold water to give thickening properties, but insoluble in hot water, where gels are formed, allowing binding of food products when heated. Films of HPMC exhibit good barrier properties.

Benefits
The hot-gelation properties are used to reduce boil-out of sauces and fillings during heating in a range of sauces and fillings. The thermal gelation properties also improve binding and hence give better shape retention in products such as reformed meats, reformed vegetables, potato products, vegetarian burgers and dietetic breads. The barrier properties can be used to reduce oil uptake in deep-fried products. HPMC has a higher gelation point and viscosity than comparable methyl cellulose types.

Limitations
Hydroxypropyl methyl cellulose is included in the group of celluloses that have been allocated a group ADI of "not specified" by JEFCA. In the EU, hydroxypropyl methyl cellulose is permitted in Part C Group 1 of Regulation 1129/2011, additives permitted at *quantum satis*.

Typical Products
Soya burgers, sausages, onion rings, formed potato products, gluten-free bakery products.

Number
E 465

Product
Methylethyl cellulose
Ethylmethyl cellulose

Sources
Methylethyl cellulose is a pale-yellowish powder produced from cellulose by treatment with alkali, dimethyl sulfate and ethyl chloride under controlled conditions, with purification by washing.

Function in Food
Methylethyl cellulose has surface activity and can stabilise foams in the presence of fat. It can also form thermoreversible gels on heating solutions.

Benefits
Solutions of methylethyl cellulose can be whipped to give good overrun. The foams are tolerant to fat and are able to stabilise egg white when fat is present.

Limitations
Methylethyl cellulose is permitted in the EU in Group 1 Part C of Regulation 1129/2011, additives permitted at *quantum satis*. It is included in the group of celluloses that have been allocated a group ADI of "not specified" by JEFCA.

Typical Products
Dairy-free creams and toppings, aerated desserts and meringues.

Number
E 466

Product
Carboxymethyl cellulose
Cellulose gum

Sources
Carboxymethyl cellulose (CMC) is manufactured from purified cellulose by reaction with monochloracetic acid under controlled conditions, followed by washing to purify it.

Function in Food
CMC is soluble in hot and cold water and is used as a thickener. It also acts as a stabiliser in frozen products. It reacts with proteins to stabilise low-pH dairy and soya products.

Benefits
CMC dissolves in both cold and hot water to give clear, flavourless solutions with a range of viscosity, depending on the choice of CMC grade. The viscosity build-up can be very rapid, especially when fine particle size products are used. This thickening function is used in a range of drinks, sauces and toppings, and in powders to be made up into these products. CMC is also widely used to stabilise fruit pulp in fruit drinks and drink concentrates. Frozen desserts such as ice creams and water ices use CMC to inhibit the growth of ice crystals and

maintain a smooth texture. CMC is also used as a water binder, especially in bakery products. It is, in fact, sodium carboxymethyl cellulose, and this ionic character leads to protein reactivity, which is used to stabilise dairy and soya products with pH level of the order of 4.5.

Limitations
CMC is permitted in the EU in Group 1 Part C of Regulation 1129/2011, additives permitted at *quantum satis*. It is included in the group of celluloses that have been allocated a group ADI of "not specified" by JEFCA.
With the finer particles size grades of CMC, the water uptake can be very rapid, leading to clumping. To avoid this, the product should either be preblended with other ingredients such as sugar, or blended with a high-shear mixer.

Typical Products
Soft drinks, powders and concentrates for drinks, ice creams and water ices, bakery products.

Number	Product
E 468	Crosslinked sodium carboxymethyl cellulose
	Crosslinked cellulose gum
	Croscarmellose

Sources
Cellulose, from wood pulp or cotton fibres, is reacted in sodium hydroxide with sodium monochloroacetate to form sodium carboxymethylcellulose which, when heated under acid conditions, crosslinks to form the crosslinked gum.

Function in Food
Crosslinked sodium carboxymethyl cellulose gum swells in water without dissolving and is used as a disintegrating agent to accelerate the break-up, dispersion and/or dissolution in water of tablets.

Limitations
Within the EU, crosslinked sodium carboxymethyl cellulose is permitted in Regulation 1129/2011 only in tablets and capsules of table-top sweeteners or food supplements to specified limits.

Typical Products
Sweetener tablets and solid dietary supplements such as vitamin, fibre and mineral tablets.

Number	Product
E 469	Enzymatically hydrolysed carboxymethyl cellulose
	Enzymatically hydrolysed cellulose gum

Sources
Enzymatically hydrolysed carboxymethyl cellulose is a white to off-white powder made by the action of a cellulase from *Trichoderma longibrachiatum*

on carboxymethyl cellulose (E 466). It is soluble in water and insoluble in ethanol.

Function in Food
This cellulose can be used as a thickener in low-fat and fat-reduced foods and in soft drinks.

Limitations
Enzymatically hydrolysed carboxymethyl cellulose is included in the group of celluloses that have been allocated a group ADI of "not specified" by JEFCA. In the EU, it is permitted in Part C Group 1 of Regulation 1129/2011, as an additive permitted at *quantum satis*.

Typical Products
There is no commercial production of this material

Number	Product
E 470a	Sodium, potassium and calcium salts of fatty acids
E 470b	Magnesium salts of fatty acids

Sources
The salts of fatty acids are made by reacting the acids with the appropriate hydroxide. The acids used are principally stearic, palmitic and oleic (see E 570). The salts can be used singly or in mixtures.

Function in Food
The salts of fatty acids have a number of uses, usually derived from their fatty acid component. Thus, they are free-flow agents and anticaking and defoaming agents.

Benefits
Magnesium stearate is used to help powders flow smoothly into moulds during tabletting. Other salts are used to decrease foam during the processing of beet sugar, as an antitack agent in chewing gum, and as a promoter of yeast activity.

Limitations
In the EU the salts are included in Group 1 of Part C in Regulation 1129/2011, additives permitted at *quantum satis*.

Typical Products
Magnesium stearate is used in food supplement tablets.

Number	Product
E 471	Mono- and diglycerides of fatty acids

Sources
Fats are compounds of glycerol and fatty acids. Glycerol has three hydroxyl groups and in fats and oils (triglycerides) all three hydroxyl groups are esterified

with a fatty acid. Mono- and diglycerides occur naturally as constituents of natural fats, and are also formed from triglycerides, as products of fat metabolism, during the digestion and absorption of food.

They are produced commercially by a) heating triglyceride fats with an excess of glycerol in the presence of, typically, a basic catalyst in a process called interesterification, or b) direct esterification of glycerol with fatty acids. The resulting composition of the mono- and diglyceride mixture is dependent upon the proportion of glycerol and the temperature conditions used. The monoester content in the mixture after removal of excess glycerol is usually in the range of 30–60%. Distillation of the mono- and diglyceride will yield a distilled monoglyceride with a purity of typically above 90%.

The composition of the product used as a food additive will vary according to conditions, but glyceryl monostearate and glyceryl distearate are often major components. To be permitted as an additive, the content of mono- and diesters must not be less than 70%, but this specification permits a wide range of compositional types for specific applications. Accordingly, these products will vary in their appearance from pale straw to brown oily liquids, to white or slightly off-white hard waxy solids. The solids may be in the form of flakes, powders or small beads. Typically, these products are insoluble in water but can form stable hydrated dispersions.

The fats, fatty acids and glycerol used as source material for mono- and diglycerides can be from either animal or vegetable origin.

Function in Food

Emulsifiers are used to disperse fat droplets in water or water droplets in fat. Because they act at the surface between the fat and the water, they are also known as surface-active agents or surfactants. Monoglycerides and mixtures of mono- and diglycerides are by far the most important commercially of all the food surfactants known; in Europe they represent no less than 50% of the total food emulsifier market and they are also important intermediates in the manufacture of the E 472 series (mono- and diacid esters of mono- and diglycerides of fatty acids) and other emulsifiers.

Mono- and diglycerides are used in bread to ensure an even texture in the final product.

Benefits

Mono- and diglycerides are used widely in a great many products.

In the production of bread, the mono- and diglycerides enable the gluten to remain plastic and pliable so that through the kneading process the strands of gluten can form a smooth extensible film, ensuring the correct texture in the finished product.

In cakes, the air bubbles in the batter are enclosed in films of protein in which the fat is dispersed. The surfactant improves the production of the initial air bubbles, ensuring their uniformity and an improved texture of the finished baked product. In margarine they have an emulsion stabilising effect, and in ice cream an emulsion de-stabilising effect, ensuring good air bubble stability.

Limitations
In the EU mono- and diglycerides are included in Group 1 Part C of Regulation 1129/2011, additives permitted at *quantum satis.*

Typical Products
Margarines and fat spreads, confectionery, ice cream, bread, cakes and baked goods, cereals, desserts.

Number	Product
E 472a	Acetic acid esters of mono- and diglycerides of fatty acids (acetems)

Sources
Acetems are made by reacting mono- and diglycerides of fatty acids with acetic anhydride (from acetic acid, E 260) or by transesterification of fats with triacetin (E 1518). They consist of glycerol combined with at least one fatty acid and one acetic acid. Acetems are available in liquid, pasty and solid forms, with a wide range of melting points with the properties dependent upon the number and nature of the fatty acids and the number of hydroxyl groups reacted with acetic acid. Acetems are insoluble in hot or cold water and dispersible or soluble in edible oils.

Function in Food
Acetems are used for aerating and foam stabilising, as moisture-resistant coatings and as lubricants in chewing gum and as release agents.

Benefits
In blends for whipped toppings or as aerating/emulsifying agents for cakes and sponges, acetems provide emulsification and stabilisation for the aqueous foams of protein, fat and sugar. In chewing gum they are used to adjust juiciness, texture and stickiness.
Acetems that are solid at room temperature are used to coat and protect foods such as sausages, fruit and cheese. Apart from preventing microbiological contamination, the barrier controls moisture migration and provides a removable surface for labelling. Films of appropriate acetems can reduce uptake of taints and extend the shelflife of products such as liver sausage by preventing contact with the air, which can lead to harmless but ugly surface discoloration. These films can also retain protective atmospheres.
In the preparation of jams and marmalades, acetems are used as antifoams to aid filling and present a neat, unbubbled surface.
Acetems are also used to manipulate the melting point and plasticity of fats.

Limitations
Within the EU, acetems are included in Group 1 Part C of Regulation 1129/2011, additives permitted at *quantum satis.*

Typical Products
Cakes, sausages and desserts.

Number   Product
E 472b   Lactic acid esters of mono- and diglycerides of fatty acids
                          (lactems)

Sources
Lactems are made by reacting mono- and diglycerides of fatty acids with lactic
acid or by reacting glycerol with a mixture of fatty acids and lactic acid (E 270)
so that the glycerol is combined with at least one fatty acid and one lactic acid.
The properties of the lactems are decided by selection of the fatty acids used to
make the glyceride backbone and the number of hydroxyl groups reacted with
lactic acid.
Lactems are available in liquid, pasty and solid forms, with a wide range of
melting points. They are yellow- to amber-coloured materials dispersible in hot
water and soluble in edible oils.

Function in Food
Lactems are used as emulsifiers in oil-in-water emulsions, frequently in
combinations with more hydrophilic emulsifiers to produce blends capable
of making stable water-in-oil emulsions. They are used to improve the incorporation and distribution of air in whipped systems such as cakes and mousses.

Benefits
Lactems are used in combination with other emulsifiers in whipped topping
concentrates, and as aerating and emulsifying agents for cakes and sponges to
produce a narrow pore-size distribution in the crumb. In mousses these
combinations of emulsifiers are used to produce consistent aeration and to
maximise volume. Lactems are also used in baking margarines and cake
shortenings.

Limitations
Within the EU, lactems are included in Group 1 Part C of Regulation
1129/2011, additives permitted at *quantum satis*.

Typical Products
Cakes and whipped toppings.

Number   Product
E 472c   Citric acid esters of mono- and diglycerides of fatty acids
                          (citrems)

Sources
Citrems are made by reacting mono- and diglycerides of fatty acids with citric
acid (E 330) or by reacting glycerol with a mixture of citric acid and fatty acids.

The product can be partly neutralised to form sodium, potassium or calcium salts.

Citrems are white to ivory coloured materials available in liquid, pasty and solid forms, with a wide range of melting points determined by the nature of the fatty acids and the proportion of hydroxyl groups combined with a citric acid. Citrems are dispersible in hot water and soluble in edible oils.

Function in Food

Citrems are used as emulsifiers to prevent separation of fat during cutting or chopping and to stabilise emulsions in cooked products such as liver sausage. They are used to reduce spattering of margarines during frying. Special citrems are used in the production of dried yeast to protect the yeast cells during drying, and in chocolate for controlling flow properties.

Benefits

Citrems are used in combination with other emulsifiers in low-calorie or fat-reduced margarines.

Limitations

Within the EU, citrems are included in Group 1 Part C of Regulation 1129/2011, additives permitted at *quantum satis*.

Because they can contain free acid groups, the surfactant ability of some citrems can be affected by pH.

Typical Products

Sausages and frying margarines, cocoa and chocolate products.

Number	Product
E 472d	Tartaric acid esters of mono- and diglycerides of fatty acids (tatems)

Sources

Tatems are made by reacting mono- and diglycerides of fatty acids with tartaric acid (E 334) in various proportions.

Tatems are white to pale yellow materials available in forms from sticky liquids to solids.

Owing to the high content of expensive tartaric acid, tatems are the most expensive of the E 472 series and because better functionality is available from datems or others, they are rarely produced commercially.

Function in Food

Tatems are used as emusifiers and stabilisers.

Limitations

Within the EU, tatems are included in Group 1 Part C of Regulation 1129/2011, additives permitted at *quantum satis*. They have been allocated an ADI of "not specified" by JEFCA.

Typical Products
None known.

Number   Product
E 472e   Mono- and diacetyl tartaric acid esters of mono- and
                           diglycerides of fatty acids (datems)

Sources
Datems are made by reacting mono- and diglycerides of fatty acids with
diacetyl tartaric acid anhydride in the presence of acetic acid (E260), or by
esterification of mono- and diglycerides with tartaric acid and acetic acid in the
presence of acetic acid anhydride. The properties of the datems are determined
by the fatty acid composition in the glyceride backbone, the relative amounts of
the raw materials, and the control of the subsequent steps of the reaction.
Datems are yellow materials available in liquid, pasty and solid forms, with a
wide range of melting points. They are dispersible in both cold and hot water.

Function in Food
Datems are used as emulsifiers in a wide range of food products, particularly
where there is potential for interaction with protein, such as in wheat-based
baked goods and egg-containing emulsions such as mayonnaise. The formation
of hydrogen bonds between the datem and the gluten proteins in wheat flour
strengthens the gluten network. Datems also stabilise egg proteins and render
them less susceptible to coagulation under the conditions of heat or shear.

Benefits
In the major use area of yeast-raised baked goods, datems work as classical
emulsifiers of oil and water to ensure even and stable distribution of lipids, but
they also improve dough performance in a number of ways. Mixing tolerance,
that is to say the length of time for which the dough can be mixed or
manipulated without overextension and loss of condition, is increased by the
use of datems. Fermentation tolerance is also improved as the period of time
for which a dough remains at or near peak of volume development is leng-
thened. The volume of baked goods can also be increased because of better gas
retention as doughs are fermented and handled.

Limitations
Within the EU, datems are included in Part C Group 1 of Regulation
1129/2011, additives permitted at *quantum satis*. In the USA, datems are
considered GRAS.
Datems must be stored in dry conditions as they hydrolyse in moist air to
release acetic acid.

Typical Products
Bread and baked goods.

Number	Product
E 472f	Mixed acetic and tartaric acid esters of mono- and diglycerides of fatty acids (matems)

Sources

Matems are made by reacting mono- and diglycerides of fatty acids (E 471) with a mixture of tartaric acid (E 334) and acetic acid (E 260) in various proportions. Since acetic acid can also react with tartaric acid to form acetylated tartaric acid, there are effectively three acids in this reaction mixture competing for the free hydroxyl groups of the fatty acid glyceride. A considerable number of different reaction products is thus possible and very tight control of the processing conditions is required to make a product of consistent quality.

Matems are water-dispersible materials ranging from white to pale yellow in colour and from sticky liquids to waxes.

Function in Food

Since matems can exhibit some of the characteristics of both acetems and datems they can be used as emulsifiers, stabilisers and antifoams. However, they are not widely manufactured or used.

Benefits

In the principal area of use, baked goods, matems work as classical emulsifiers of oil and water but can also interact with gluten proteins. They improve the tolerance of doughs to extended mixing and increase the period time for which a dough remains at or near the peak of volume development.

Limitations

Within the EU, matems are included in Group 1 Part C of Regulation 1129/2011, additives permitted at *quantum satis*.

Matems must be stored in the dry as they react with moist air to release acetic acid.

Typical Products
Baked goods

Number	Product
E 473	Sucrose esters of fatty acids

Sources

Sucrose esters of fatty acids are prepared by esterifying one or more of the (primary) hydroxyl groups of sucrose with the methyl and ethyl esters of food fatty acids or by extraction from sucroglycerides (E 474). Depending upon the degree of esterification, a wide range of sucrose esters is obtained, covering the major part of the hydrophile–lipophile (HLB) scale.

Products made from unsaturated fatty acids are yellowish materials, pasty or waxy in texture, while those from saturated fatty acids are white to greyish powders. The esters are sparingly soluble in water.

Function in Food
Sucrose esters are generally used as emulsifiers in oil-in-water emulsions, but can also be used as texturisers in fine bakery wares or to stabilise foam in dairy products and their analogues.
The sucrose esters exhibit some antimicrobial activity, especially against Gram-positive bacteria.

Benefits
Sucrose esters are neutral in taste and odour and do not influence the taste of other ingredients present in a formulation. Being heat stable, heating to temperatures up to 185 °C is possible without any negative effect on performance. They are stable to a range of pH levels from 4 to 8. Sucrose esters are efficient surfactants and can be used at low dosage levels.

Limitations
Sucrose esters are permitted in the EU in Regulation 1129/2011 in a range of products with individual maxima. They have been allocated an ADI of 0–30 mg/kg body weight by JEFCA.

Typical Products
Ice cream, fine bakery wares, cream analogues, powders for hot beverages.

Number	Product
E 474	Sucroglycerides

Sources
Sucroglycerides are manufactured by reacting sucrose with edible fats or oils to produce a mixture of mono-, di- or triesters of sucrose with mono-, di- and triglycerides. The fats are generally commercial vegetable fats such as palm oil, rapeseed oil, coconut oil or castor oil.
Sucroglycerides are white to off-white gels or powders that are insoluble in cold water.

Function in Food
Sucroglycerides are emulsifiers and dispersing aids and help to control crystallisation and improve texture.

Benefits
Sucroglycerides are less expensive than sucrose esters. The different types of triglycerides used in the manufacture allow development of a wide range of products with a range of functionalities; however, they are rarely made or used.

Limitations
Sucroglycerides are permitted in the EU in Regulation 1129/2011 in a range of products with individual maxima. They have been allocated an ADI of 0–30 mg/kg body weight by JEFCA.

Typical Products
Nonalcoholic drinks, bakery products and ice cream.

Number	Product
E 475	Polyglycerol esters of fatty acids

Sources
Polyglycerol esters of fatty acids are made by reaction of polyglycerol with one or more fatty acids. Polyglycerol is made from glycerol (E 422) by heating it under vacuum with a catalyst to make a mixture of predominantly di-, tri- and tetraglycerol.
The polyglycerol esters are light- to amber-coloured materials ranging from oily to a hard waxy consistency and dispersible in water and soluble in edible oils.

Function in Food
Polyglycerol esters of fatty acids are more polar than mono- and diglycerides so they are used to hold water in fat emulsions. They are also used to facilitate aeration in cake mixes.

Benefits
The esters act synergistically with other emulsifiers to reduce the total amount of emulsifier required in products such as desserts and to reduce the amount of hard fat needed in margarine blends. They also reduce the tendency of low-fat spreads to weep on storage.

Limitations
Polyglycerol esters are permitted in the EU in Part C Group 1 Regulation 1129/2011 in a range of products with defined limits.
They have been allocated an ADI of 0–25 mg/kg body weight by JECFA.

Typical Products
Cakes and gateaux.

Number	Product
E 476	Polyglycerol polyricinoleate

Sources
Polyglycerol polyricinoleate (PGPR) is a mixture of partial esters of poly-glycerol with linearly interesterified castor oil (*ricinus communis*) fatty acids (ricinoleic acid) made by esterifying castor oil fatty acids with polyglycerol (E 475). The polyglycerol moiety is predominantly di-, tri- and tetraglycerol. Condensed castor oil fatty acids are made by heating castor oil fatty acids in an

inert atmosphere and condensing to an average of five fatty acid residues per molecule. PGPR is a viscous light brown liquid at room temperature that is soluble in edible oils.

Function in Food
PGPR is used to modify the flow characteristics of chocolate and, because it is an efficient surfactant, to stabilise water-in-oil emulsions such as low-fat spreads.

Benefits
PGPR acts as a viscosity modifier in chocolate- and cocoa-based products. Rheology in chocolate is complex, and the flow behaviour is described by two parameters – yield value and plastic viscosity. Yield value relates to the force needed to start liquid chocolate moving and plastic viscosity to the force necessary to keep it moving.
PGPR decreases yield value and therefore improves the flow properties and handling of chocolate for coating, moulding and block chocolate production. It also decreases the risks of defects such as air bubbles in the finished product. PGPR also has a synergistic effect with lecithin, which has a beneficial influence on plastic viscosity. Use of PGPR allows the reduction of fat levels in the food product.

Limitations
In the EU, PGPR is permitted in Regulation 1129/2011 only in low or very low-fat spreads and in cocoa- and chocolate-based confectionery. JECFA has allocated an ADI of 0–7.5 mg/kg body weight.

Typical Products
Chocolate, chocolate products and low-fat spreads.

Number	Product
E 477	Propane 1,2 diol esters of fatty acids

Sources
Propane 1,2 diol esters of fatty acids are made by reaction of the acids with 1,2 epoxypropane or by reaction of propylene glycol (propane 1,2 diol) with oils such as soya bean oil. Propane 1,2 diol esters of fatty acids are white to yellow materials dispersible in hot water and soluble in edible oils. Typically the fatty acids are palmitic and stearic and the process produces a mixture of mono- and diesters, which can be distilled to produce up to 90% monoester.

Function in Food
The esters are emulsifiers used to improve the whippability of powdered desserts and the texture and volume of cakes.

Benefits
The propane 1,2 diol esters are synergistic with other emulsifiers such as the mono- and diglycerides of fatty acids (E 471).

Limitations

The propane 1,2 diol esters are permitted in the EU in Regulation 1129/2011 in a range of products with specific limits in each case.

JECFA has allocated an ADI of 0–25 mg/kg body weight calculated as propylene glycol.

Typical Products

Powdered desserts.

Number	Product
E 479b	Thermally oxidised soya-bean oil interacted with mono- and diglycerides of fatty acids.

Sources

As the name suggests, the product is made by first heating soya oil in the presence of air at 190–200 °C followed by reaction with mono- and diglycerides of fatty acids (E 471). It consists of a triglyceride with all hydroxyl groups esterified by a fatty acid, originating from either the soya oil or from the mono- and diglycerides.

The product has a light-brown colour with a waxy to solid consistency. It is dispersible in hot water and soluble in edible oils.

Function in Food

The mixture is used to stabilise fat emulsions used for frying on a hot plate or griddle.

Benefits

E 479b provides a stable emulsion that has low viscosity but coats the frying surface, does not char during frying and gives good release of the product from the griddle, leaving it clean.

Limitations

E 479b is permitted in the EU in Regulation 1129/2011 only for use in fat emulsions for frying purposes.

Number	Product
E 481	Sodium stearoyl-2-lactylate
E 482	Calcium stearoyl-2-lactylate

Sources

Stearoyl lactylates are white- to ivory-coloured waxy materials made by reacting stearic acid and lactic acid together and neutralising the resulting acid with the appropriate base. They are dispersible in hot water and soluble in hot edible oils.

Function in Food

The stearoyl lactylates are hydrophilic emulsifiers that are used to disperse fat evenly in water-based formulations.

Benefits
The lactylates are used to distribute fat in bread dough to give a uniform crumb structure and improve keeping quality. They are also used in emulsifier blends to improve fat suspension during spray drying of fat powders such as beverage whiteners. They are ionic and bind to both proteins and starches.
The calcium salt in particular is used to increase strength and volume in bread and to increase the tolerance to processing.

Limitations
The lactylates have been allocated an ADI of 0–20 mg/kg body weight by JECFA. In the EU they are permitted in Regulation 1129/2011 in a range of products with individual limits.

Typical Products
Bread, beverage whiteners and low-fat spreads.

Number	Product
E 483	Stearyl tartrate

Sources
Stearyl tartrate is made by reacting commercial stearyl alcohol with tartaric acid (E 334). Commercial stearyl alcohol is actually a mixture of stearyl and palmityl alcohols and E 483 is a mixture of distearyl tartrate, dipalmityl tartrate and stearyl palmityl tartrate.

Function in Food
Stearyl tartrate is used as an emulsifier.

Limitations
Within the EU stearyl tartrate is permitted in Regulation 1129/2011 in bakery wares to a maximum of 4 mg/kg and in desserts to a maximum of 5 g/kg.

Typical Products
None known.

Number	Product
E 491	Sorbitan monostearate
E 492	Sorbitan tristearate
E 493	Sorbitan monolaurate
E 494	Sorbitan monooleate
E 495	Sorbitan monopalmitate

Sources
The sorbitan esters are produced by the reaction of the appropriate fatty acid with hexitol anhydride, which is itself derived from sorbitol (E 420).

Function in Food

The sorbitan esters are nonionic emulsifiers, which have a range of hydrophile–lipophile balance (HLB) values from 2 to 8. The HLB system indicates whether emulsifiers will tend to favour oil-in-water or water-in-oil emulsions, and a figure of lower than 6 indicates a preference towards water-in-oil. The tristearate is particularly lipophilic.

Benefits

The sorbitan esters are used to hold aqueous solutions in suspension in fatty materials. Thus, they are used to disperse aqueous additives in ice cream, fat spreads and desserts, to provide stable emulsions of fat for spray drying, as beverage whiteners or fat powders, and to inhibit staling in bakery products by interrupting the structure. They are also used as antifoaming agents in the product of beet sugar, boiled sweets and preserves, and to modify fat crystal structure in chocolate to inhibit the development of the storage defect known as "bloom".

The esters tend to be used in mixtures, with each other or with the polyoxy-ethylene sorbitan esters (E 432 to E 435) to generate the optimum HLB value for the particular food system.

Limitations

In the EU, sorbitan esters are included in Regulation 1129/2011 where they are permitted in a range of products with individual maxima in each case.

They have been allocated an ADI of 0–25 mg/kg body weight by JECFA.

Typical Products

Fat spreads and cake mixes.

Number	Product
E 500	Sodium carbonates
	(i) Sodium carbonate
	(ii) Sodium hydrogen carbonate (sodium bicarbonate)
	(iii) Sodium sesquicarbonate

Sources

Sodium bicarbonate is made industrially from brine and limestone using the ammonia soda process. It is purified by repeated crystallisation.

Sodium carbonate is made by heating the impure sodium bicarbonate. It is also produced in the USA from sodium sesquicarbonate ore by heating, followed by leaching with warm water.

Sodium sesquicarbonate is mined in the USA, where it is known as "trona". It is also produced by crystallising a mixture of sodium carbonate and sodium bicarbonate.

Function in Food
Of the carbonates, the most common is sodium bicarbonate. It is also known as bicarbonate of soda or baking soda. It is used in baking powder, to generate carbon dioxide by mixing it with an acidic material such as tartaric acid (E 334) or one of the acidic phosphates. Sodium bicarbonate does decompose thermally and can be used alone as a raising agent.

Sodium carbonate is also used as a raising agent in cakes, in combination with, for example, disodium diphosphate (E 450(i)).

The sodium carbonates are also used to modify the acidity of products and to stop the hydrolysis reaction in the production of invert sugar.

Benefits
The sodium carbonates are soluble in cold water, readily available and inexpensive.

Their rate of reaction with acids can be varied by changing the particle size and both carbonate and bicarbonate are available in a number of granular sizes. They are also available as granules coated with fat for applications where the reaction needs to be inhibited until later in the process.

The bicarbonate is also used alone, generating carbon dioxide by the action of heat at temperatures as low as 60 °C.

Limitations
When used in excess the bicarbonate can leave a soapy taste in the product. The particle size of the carbonate has to be chosen with care since use of too large a particle can result in there being unreacted carbonate in the final product.

In the EU the carbonates are included in Group 1 of Part C in Regulation 1129/2011, additives permitted at *quantum satis*.

Typical Products
Baked goods including pastries, cakes, waffles, cookies and scones.

Number	Product
E 501	Potassium carbonates
	(i) Potassium carbonate
	(ii) Potassium hydrogen carbonate

Sources
Both potassium carbonate and bicarbonate are prepared by passing carbon dioxide into potassium hydroxide. The carbonate is formed by adding more carbon dioxide after the formation of the bicarbonate.

Function in Food
Potassium carbonate is a raising agent in conjunction with an acidic material such as disodium diphosphate (E 450(i)). It is also used in the alkalisation of cocoa powder.

Potassium bicarbonate can also be used in baking powder to generate carbon dioxide by mixing it with an acidic material such as tartaric acid (E 334) or disodium diphosphate (E 450(i)).

The bicarbonate is also used alone, generating carbon dioxide by the action of heat at temperatures as low as 60 °C.

Benefits

The potassium carbonates are used as raising agents where it is necessary to restrict the amount of sodium or enhance the potassium in the product. Potassium carbonate is more soluble than sodium carbonate.

In the alkalisation of cocoa powder, the powder is reacted with an alkali to deepen the colour and increase the intensity of flavour. A number of alkalis are used for this purpose, each having its particular advantages. Potassium carbonate is considered to give a better colour than sodium carbonate.

Limitations

Potassium carbonate only releases carbon dioxide when used in conjunction with an acid. It is thus less convenient to use than sodium carbonate.

When used in excess, the bicarbonate can leave a soapy taste in the product.

Potassium bicarbonate is more expensive and requires higher usage rates than sodium bicarbonate.

Potassium carbonates are included in Group 1 Part C of Regulation in 1129/2011, additives permitted at *quantum satis*.

Typical Products

Low-sodium crackers, biscuits and energy bars, cocoa powder and chocolate drinks.

Number	Product
E 503	Ammonium carbonate
	(i) ammonium carbonate
	(ii) ammonium bicarbonate

Sources

Ammonium bicarbonate is prepared by passing carbon dioxide into ammonia solution. The bicarbonate precipitates and is washed and dried. The carbonate is prepared by subliming a mixture of ammonium sulfate and calcium carbonate together and is actually a mixture of the bicarbonate and the carbamate.

Function in Food

The ammonium carbonates are frequently used as mixtures as they have similar performance. Both are used in baking to generate carbon dioxide by the action of heat or with acids. They can also be used in the alkalising of cocoa powder.

Benefits
The ammonium carbonates are particularly useful because they break up on heating to only 60 °C, generating both carbon dioxide and ammonia, and leaving no residue in the product. They are both readily soluble in water.

Limitations
Ammonium carbonates tend to be used only in thin products with final moisture content below 5%, such as biscuits. Products outside these constraints, such as cakes or soft cookies, can retain ammonia with the final product with an adverse impact on quality. The carbonate must be stored in sealed containers as it loses ammonia and carbon dioxide on exposure to the air, to leave the bicarbonate. Both products must be stored at or below room temperature. In the EU, ammonium carbonates are included in Group1 Part C of Regulation 1129/2011, additives permitted at *quantum satis*.

Typical Products
Biscuits, crackers and sugar confectionery.

Number	Product
E 504	Magnesium carbonates
	(i) magnesium carbonate
	(ii) magnesium hydrogen carbonate (magnesium bicarbonate)

Sources
Magnesium carbonate is made from dolomite, a naturally occurring mineral. Magnesium bicarbonate is made by passing carbon dioxide into a magnesium hydroxide slurry at high pressure.

Function in Food
Magnesium carbonate is used as a source of carbon dioxide, either using heat or acid. It is also a source of magnesium in fortified products and a free-flow agent in salt.
It has been used as a pharmaceutical antacid and can be used to reduce the acidity of foodstuffs.

Benefits
Magnesium carbonate is inexpensive, available as a fine powder and is not hygroscopic, all of which make it useful as a free-flow agent.

Limitations
Magnesium carbonate is included in Group 1 Part C of Regulation 1129/2011 in the EU, additives permitted at *quantum satis*.

Typical Products
Cheese, ice cream and in table salt.

Number   Product
E507   Hydrochloric acid

Sources
Hydrochloric acid is made industrially from salt.

Function in Food
Hydrochloric acid is a strong acid and is used to increase the acidity of formulations and in the hydrolysis of large molecules.

Benefits
Hydrochloric acid is a strong acid and is very cost effective. It is often used because it has a less acid taste than other acids, such as citric acid (E 330).
It is also used in the production of invert sugar from sucrose and in the production of glucose syrups from starch.
It is used in the hydrolysis of vegetable proteins and has the advantage that the usual product of neutralisation is salt, which enhances the taste of savoury products such as the hydrolysed proteins.

Limitations
Concentrated hydrochloric acid is corrosive and it must be handled with great care to avoid contact with skin. Once it is diluted in the food it is harmless. Because it is a strong acid it usually needs to be mixed in rapidly to avoid local decreases in pH that could have irreversible effects.
The degree of hydrolysis of starch that can be achieved with hydrochloric acid is limited and enzymes are used to produce high sweetness syrups.
In the EU, hydrochloric acid is included in Group 1 Part C of Regulation 1129/2011, additives permitted at *quantum satis*.

Typical Products
Invert (golden) syrup, glucose syrup, hydrolysed vegetable proteins.

Number   Product
E508   Potassium chloride

Sources
Potassium chloride is purified from a natural mineral source.

Function in Food
Potassium chloride has a taste similar to salt and is used to provide saltiness in products where low sodium content is required. It is also used on its own in table-top salt replacers for people on low-salt diets.

Benefits
Potassium chloride is used in products for people who wish to limit their sodium intake as it contains no sodium ions.

Limitations
Potassium chloride does not taste the same as common salt (sodium chloride) and is not a complete replacement.
Within the EU, potassium chloride is included in Group1 Part C of Regulation 1129/2011, additives permitted at *quantum satis*.

Typical Products
Table-top salt replacers and dietetic foods.

Number	Product
E509	Calcium chloride

Sources
Calcium chloride is extracted from natural brines or manufactured as a byproduct of the production of sodium carbonate.

Function in Food
Calcium chloride is used as an aqueous solution to provide a source of calcium ions.

Benefits
Calcium chloride is soluble in water which makes it a good source of calcium ions in solution. These ions are used for a number of purposes depending on the product. In brewing they modify the hardness of the water, in canned vegetables and vegetable products, they improve texture by reacting with the natural pectin and they can also be used to crosslink alginate gels.
Calcium chloride is also used to aid coagulation in cheese manufacture and in the extraction of alginates from seaweed.

Limitations
In the EU calcium chloride is included in Group 1 Part C of Regulation 1129/2011, additives permitted at *quantum satis*.

Typical Products
Canned and bottled fruit and vegetables such as carrots, kidney beans, gherkins, olives and pickles, and ketchup, and in cheese.

Number	Product
E 511	Magnesium chloride

Sources
Magnesium chloride is available as a mineral ore, from underground brines, and is made from the natural mineral dolomite by reaction with hydrochloric acid (E 507).

Function in Food
Magnesium chloride is used in the preparation of water for brewing and as a source of magnesium in fortified products.

Benefits
Magnesium chloride is inexpensive and readily soluble in water.

Limitations
In the EU, magnesium chloride is included in Group 1 Part C in Regulation 1129/2011, additives permitted at *quantum satis*.

Typical Products
None known.

Number	Product
E512	Stannous chloride

Sources
Stannous chloride is made by reacting tin with either chlorine or hydrochloric acid under the appropriate conditions.

Function in Food
Stannous chloride is used to maintain the colour of processed asparagus.

Limitations
In the EU stannous chloride is only permitted in Regulation 1129/2011 in canned and bottled white asparagus to a maximum level of 25 mg/kg.
Stannous chloride absorbs oxygen from the air and should be kept in a tightly sealed container in a cool place.

Number	Product
E513	Sulfuric acid

Sources
Sulfuric acid is made industrially from sulfur dioxide (E 220).

Function in Food
Sulfuric acid is a strong acid and is used to increase the acidity of formulations. It is also used in the production of invert sugar.

Benefits
Sulfuric acid is a strong acid and is very cost effective. The salts formed on neutralising it with common bases have little flavour.

Limitations
Concentrated sulfuric acid is corrosive and it must be handled with great care to avoid contact with the skin. Once it is diluted in the food it is harmless.

Because it is a strong acid, it usually needs to be mixed in rapidly to avoid local decreases in pH that could have irreversible effects.

In the EU, sulfuric acid is included in Group1 Part C of Regulation 1129/2011, additives permitted at *quantum satis*.

Typical Products
None known.

Number	Product
E514	Sodium sulfates
	(i) sodium sulfate
	(ii) sodium hydrogen sulfate (sodium bisulfate)

Sources
Sodium sulfate is produced as a byproduct of a number of processes using sulfuric acid. It is also mined as Glauber's salt and purified by recrystallisation.

The bisulfate is made by further reaction of the sulfate with sulfuric acid.

Function in Food
Sodium sulfate is used in colours to standardise the colour strength of the powder.

Sodium hydrogen sulfate is used as an acid in raising agents.

Benefits
Colours, whether natural, nature-identical or synthetic, do not have exactly the same intensity in every batch. Sodium sulfate is used as a neutral material to be blended with the colour to ensure that batches as sold are of consistent intensity.

Limitations
The sodium sulfates are included in Group 1 Part C, additives permitted at *quantum satis*, in Regulation 1129/2011, in the EU.

Typical Products
Powdered food colours.

Number	Product
E515	Potassium sulfates
	(i) potassium sulfate
	(ii) potassium hydrogen sulfate (potassium bisulfate)

Sources
Potassium sulfates are made by partial or complete neutralisation of sulfuric acid by potassium hydroxide, or by reaction of the acid with potassium chloride.

Function in Food
Potassium bisulfate is used as an acidic material in raising agents. It is used as a replacement for sodium bisulfate in products where it is required to reduce the sodium level.

Limitations
The potassium sulfates are included in Group 1 Part C, additives permitted at *quantum satis*, in Regulation 1129/2011, in the EU.

Typical Products
None known.

Number	Product
E 516	Calcium sulfate

Sources
Calcium sulfate is a naturally occurring mineral, also known as Plaster of Paris. It is also a byproduct of a number of manufacturing processes. Both anhydrous and hydrated forms are available.

Function in Food
Calcium sulfate is used to provide a source of calcium ions.

Benefits
Calcium sulfate is used in the preparation of water for brewing to provide both calcium and sulfate ions, which are present in naturally hard water. In canned fruit and vegetables it is also used to provide calcium ions for reaction with natural cell wall pectin to maintain the firmness of the pieces. In baking, it helps bubble stability and cell strength.

Limitations
Calcium sulfate is barely soluble in water.
In the EU it is included in Group 1 Part C in Regulation 1129/2011, additives permitted at *quantum satis*.

Typical Products
Wafer biscuits, bread, beer, canned fruit and vegetables and tabletted products.

Number	Product
E 517	Ammonium sulfate

Sources
Ammonium sulfate is a white solid produced by passing ammonia gas into sulfuric acid solution.

Function in Food
Ammonium sulfate is used as a yeast nutrient and as a carrier for flavours.

Limitations
Within the EU, ammonium sulfate is only permitted in Regulation 1130/2011
as a carrier in food additives and food enzymes.

Number	Product
E 520	Aluminium sulfate
E 521	Aluminium sodium sulfate
E 522	Aluminium potassium sulfate
E 523	Aluminium ammonium sulfate

Sources
Aluminium sulfate is manufactured by the reaction of sulfuric acid with
naturally occurring aluminium oxide or as a byproduct in the manufacture of
alcohols.
The mixed sulfates are prepared by mixing concentrated solutions of the two
components and allowing them to crystallise as they cool. All the mixed salts
are known as "alums", although the original alum is the potassium salt.

Function in Food
Aluminium sulfate is used to improve the resistance of the conalbumin fraction
of egg white to denaturation during heat treatment, and thus to preserve the
whipping properties of dried egg white. It is not in common use in the EU.
In some countries of the world sodium aluminium sulfate is used as a component
of baking powder and for maintaining the crispness of vegetables in pickles.
It is commonly used at the flocculation stage of drinking-water treatment
plants.

Limitations
In the EU the aluminium sulfates are permitted in Regulation 1129/2011 only
in crystallised fruit to a maximum of 200 mg/kg and in egg white to a maximum
of 30 mg/kg.

Number	Product
E 524	Sodium hydroxide

Sources
Sodium hydroxide is manufactured industrially by the electrolysis of salt.

Function in Food
Sodium hydroxide is strongly alkaline and is used to decrease the acidity (raise
the pH) of food formulations.

Benefits
Because it is a strong base, sodium hydroxide is very cost effective and is used at
very low levels. It is used to neutralise acids and to stop the reaction in the

production of invert sugar. It is also used in the alkalisation of cocoa powder and in the hydrolysis of proteins.

It is used in potato processing to improve the efficiency of peeling.

Limitations

Sodium hydroxide is available as both a solid (pellets or flakes) and a concentrated liquid, both of which are very caustic, and precautions need to be taken to avoid these materials coming into contact with the skin. Once diluted in the food it is harmless. Because it is a strong base, great care needs to be taken in its use to avoid severe local increases in pH, which could have irreversible effects.

Sodium hydroxide should be kept in sealed containers because it absorbs water and carbon dioxide from the atmosphere.

Sodium hydroxide is included in Group 1 Part C in Regulation 1129/2011 in the EU, additives permitted at *quantum satis*.

Typical Products

Jams, milk drinks and cocoa powder.

Number	Product
E525	Potassium hydroxide

Sources

Potassium hydroxide is manufactured industrially by electrolysis of naturally occurring potassium chloride.

Function in Food

Potassium hydroxide is a strong base and is used to reduce acidity (increase pH) in foods.

Benefits

Potassium hydroxide is more expensive than sodium hydroxide but is used instead of sodium hydroxide in formulations where it is important to limit the amount of sodium in the final product.

Limitations

Because it is a strong base great care needs to be taken to ensure that it is mixed rapidly into the formulation to avoid local increases in pH which could have irreversible effects. Potassium hydroxide is available as both a solid and a concentrated liquid, both of which are very caustic and precautions need to be taken to avoid these materials coming into contact with skin. Once diluted in the food it is harmless.

In the EU, potassium hydroxide is included in Part C Group 1 of Regulations 1129/2011, additives permitted at *quantum satis*.

Typical Products

None known.

Number	Product
E526	Calcium hydroxide

Sources
Calcium hydroxide is made industrially by adding water to calcium oxide. It is commonly known as slaked lime. It is often made on site as part of the manufacturing process in which it is used.

Function in Food
Calcium hydroxide is a weak alkali and is used to lower acidity.
It is used in the purification of sugar syrup.

Benefits
Calcium hydroxide is used in purification of sugar because it neutralises the sugar syrup and is then removed by passing carbon dioxide into the mix, precipitating calcium carbonate and removing with it many of the organic colloidal impurities in the sugar.
It is used as a source of calcium ions to react with natural pectins in fruit to preserve the integrity of fruit particles in fruit pulp for jam manufacture.

Limitations
Calcium hydroxide is not very soluble in water.
In the EU, it is included in Part C Group 1 of Regulation 1129/2011, additives permitted at *quantum satis*.

Typical Products
Fruit fillings.

Number	Product
E 527	Ammonium hydroxide

Sources
Ammonium hydroxide is prepared by passing ammonia gas into water.

Function in Food
Ammonium hydroxide is an alkali and is used to decrease the acidity of food formulations.

Limitations
Ammonium hydroxide is available only as a solution in water.
In the EU, it is included in Part C Group1 of Regulation 1129/2011, additives permitted at *quantum satis*.

Typical Products
None known.

Number	Product
E 528	Magnesium hydroxide

Sources
Magnesium hydroxide is manufactured from the natural mineral ore dolomite by heating and hydration, or is extracted from seawater.

Function in Food
Magnesium hydroxide has long been used as a pharmaceutical antacid, and in foods is used to reduce the acidity of products.

Benefits
It provides a source of magnesium, which is an essential mineral.

Limitations
In the EU, magnesium hydroxide is included in Group 1 Part C of Regulation 1129/2011, additives permitted at *quantum satis*.

Typical Products
Fortified foods.

Number	Product
E 529	Calcium oxide

Sources
Calcium oxide is made by heating limestone

Function in Food
Calcium oxide is used as a source of calcium hydroxide (E 526). It is also used as a dough conditioner in bread making and in the production of maize tortillas.

Limitations
Calcium oxide is included in Part C Group 1 of Regulation 1129/2011 in the EU, additives permitted at *quantum satis*.

Typical Products
Fortified foods.

Number	Product
E 530	Magnesium oxide

Sources
Magnesium oxide is made industrially by heating naturally occurring magnesium carbonate (dolomite).

Function in Food
Magnesium oxide is used as a source of magnesium hydroxide (E 528).

Limitations
Magnesium oxide is included in Part C Group1 of Regulation 1129/2011 in the EU, additives permitted at *quantum satis*.

Typical Products
Fortified foods.

Number	Product
E 535	Sodium ferrocyanide
E 536	Potassium ferrocyanide
E 538	Calcium ferrocyanide

Sources
The ferrocyanides are made by reacting the respective metal cyanide with ferrous sulfate.

Function in Food
The ferrocyanides, particularly the sodium salt, are used as anticaking agents in table salt.

Limitations
The ferrocyanides are included in Regulation 1129/2011 in the EU where they are only permitted as an additive in salt and its substitutes and then only to a maximum level of 20 mg/kg.
The ferrocyanides have been allocated a joint ADI of 0.025 mg/kg body weight.

Number	Product
E 541	Sodium aluminium phosphate

Sources
Sodium aluminium phosphate is a white, odourless powder made by the reaction of sodium hydroxide, aluminium oxide and phosphoric acid.

Function in Food
Sodium aluminium phosphate is an acidic product used as a raising agent with a carbon dioxide generator such as sodium bicarbonate (E 500). It provides slow release of carbon dioxide and is used in commercial doughs and batters, where the dough is made up and is held refrigerated before cooking. Typically, about 20% of available carbon dioxide is released from the bicarbonate during mixing and the remainder is released during cooking.
It is also used in a mixture with monocalcium phosphate (E 341) to provide release of carbon dioxide both before and during cooking.

Benefits
Sodium aluminium phosphate has a bland flavour and provides a uniform texture with a large bake-out volume. Its particular benefit is the low level of carbon dioxide released during refrigerated storage.

Limitations
In the EU, sodium aluminium phosphate is permitted in Regulation 1129/2011 only in scones and sponge wares to a maximum of 1000 mg/kg.

Typical Products
Scones.

Number	Product
E 551	Silicon dioxide

Sources
Silicon dioxide (silica) is found in nature as sand. The products used in the food industry are synthetically produced amorphous silica. Two forms are available: silica aerogel, which is a microcellular silica, and hydrated silica, which is prepared by precipitation or gelling.

Function in Food
Food-grade silicon dioxide is an extremely fine powder with a very high ratio of surface area to weight. A range of surface area/weight ratios is available.
This property allows the silicon dioxide particles to coat the surface of powders to prevent them from sticking. The property also allows its use as a carrier and occasionally a thickener.

Benefits
Silicon dioxide is used at very low levels to improve powder flow and processing. It can be used to improve the performance of powders where the problem is caused by particle-size distribution, fattiness or stickiness. Particular uses include improving the flow of powders into continuous flow mixers to improve the consistency of the mix.

Limitations
The very small particle size of many grades of silicon dioxide mean that they are very dusty and care must be taken when handling them. In the EU, in Regulation 1129/2011, silicon dioxide is limited alone or in combination with silicates to use in a range of products with individual limits in each case. In Regulation 1130/2011 it is permitted at *quantum satis* as a carrier for emulsifiers and colours and with limits for dry preparations of polyols and emulsifiers.

Typical Products
Powders for drinks and desserts.

Number   Product
E 552   Calcium silicate

Sources
Calcium silicate is a white, amorphous, powder prepared by a precipitation process from inorganic raw materials.

Function in Food
Calcium silicate is used to help powders flow more freely during food processing and as an antitack agent to coat products that have a propensity to bond together.

Limitations
Within the EU, calcium silicate is controlled with silicon dioxide and other silicates in Regulation 1129/2011 where they are limited to use in a range of products with individual limits in each case. In Regulation 1130/2011 it is permitted at *quantum satis* as a carrier for emulsifiers and colours and with limits for dry preparations of polyols and emulsifiers.

Typical Products
Sliced cheese and powdered foods.

Number   Product
E 553a   Magnesium silicate

Sources
Magnesium silicate is a white amorphous powder prepared by a precipitation process from inorganic raw materials.

Function in Food
Magnesium silicate is used as a free-flow agent to improve the flow properties of powders either during processing or in use by the consumer.

Limitations
Magnesium silicate is controlled in the EU together with silicon dioxide and other silicates in Regulations 1129/2011 where they are permitted in a range of products with individual limits in each case.

Typical Products
Powdered food ingredients.

Number   Product
E 553b   Talc

Sources
Talc is a white to greyish powder and is a naturally occurring form of magnesium silicate. Sources known to be associated with asbestiform minerals are not used as food grade.

Function in Food
Talc is used as a dusting powder, anticaking agent and release agent. It is also used as a filtration aid.

Benefits
Talc is readily available, easy to handle and relatively inexpensive. Because it is relatively coarse it is easier to handle than many anticaking agents and is used for coating products such as sausages to prevent them sticking together.

Limitations
Talc is one of the free-flow agents controlled in the EU through Regulation 1129/2011. In conjunction with silicon dioxide and other silicates, it is permitted in a range of products with individual limits in each case. It is permitted in Regulation 1130/2011 up to 50 mg/kg as a carrier for colours.

Typical Products
Tablet coatings.

Number	Product
E 554	Sodium aluminium silicate
E 555	Potassium aluminium silicate
E 556	Calcium aluminium silicate

Sources
There is no commercial preparation of calcium aluminium silicate. The other two aluminium silicates are prepared by coprecipitation of soluble salts of aluminium and the appropriate metal. They are white, amorphous powders.

Function in Food
The aluminium silicates are fine powders that are used as free-flow agents in food powders. The use of sodium aluminium silicate far exceeds that of the potassium salt.
Potassium aluminium silicate is also used as a carrier.

Benefits
The aluminium silicates are relatively inexpensive. Choices between free-flow agents will be affected by both their effectiveness in improving powder flow and the effect on subsequent processing, as well as price. Potassium aluminium silicate is used where it is required to minimise the sodium content of the foodstuff.

Limitations
The silicates are included in the group of silicates controlled by Regulation 1129/2011 in the EU. In conjunction with silicon dioxide and other

silicates they are permitted in a range of products with individual limits in each case.

Potassium aluminium silicate is also permitted in Regulation 1130/2011 as a carrier for the colours titanium dioxide (E 171) and iron oxides and hydroxides (E 172). Under Commission Regulation 380/2012 calcium aluminium silicate will not be permitted to be added to foods in the EU after 31 January 2012.

Typical Products
Powders for drinks and desserts.

Number	Product
E 558	Bentonite

Sources
Bentonite is a naturally occurring clay, an aluminium magnesium silicate, formed by the weathering of volcanic ash, which is extracted and then cleaned by washing before drying and milling.

Function in Food
The major uses for bentonite are outside the food industry. It has been used as a carrier for colours but is no longer used for this purpose.

Benefits
Bentonite is an extremely absorbent material that can be prepared with a high surface area per gram.

Limitations
Under Commission Regulation 380/2012 bentonite is only permitted in the EU until 31st May 2013.

Typical Products
None known.

Number	Product
E 559	Kaolin
	Aluminium silicate

Sources
Kaolin is a naturally occurring white clay that is mined and washed before drying and milling.

Function in Food
Kaolin is a fine powder used as a dusting powder, anticaking agent and release agent. It is also used as a filtration aid.
Its major uses are outside the food industry.

Benefits
Kaolin is readily available, easy to handle and relatively inexpensive.

Limitations
Under Commission Regulation 380/2012 kaolin is only permitted in the EU until 31st January 2014.

Typical Products
None known.

Number	Product
E 570	Fatty acids

Sources
The fatty acids include stearic, palmitic and oleic acids. They are made by fractionation of natural fats such as tallow followed by acidification. They can be used alone or in combination.

Function in Food
Fatty acids have a number of functions, including as plasticisers for chewing gum and as antifoaming agents for jams.

Limitations
Fatty acids are included in Group 1 of Part C of Regulation 1129/2011, additives permitted at *quantum satis*.

Typical Products
None known.

Number	Product
E 574	Gluconic acid

Sources
Gluconic acid is an organic acid occurring naturally in plants, fruits and other foodstuffs such as wine and honey. The material of commerce is prepared by oxidation of glucose in a fermentation process; the glucose itself being produced by enzymic hydrolysis of starch. The process results in a mixture of both gluconic acid and glucona-delta-lactone (E 575). The two products are separated by the crystallisation of the lactone. The acid is supplied as a 50% solution since the dehydration leads to the formation of the lactone.

Function in Food
Gluconic acid is used as a mild acid and as a chelating agent, particularly in foods with neutral pH. However, its role as a metal chelation agent finds its major applications outside the food industry.

Benefits
The use of gluconic acid allows the reduction of pH in foods where the acid taste of citric, malic or even lactic acid would not be acceptable. The chelating effect is used to bind metal ions that might otherwise catalyse oxidation reactions that would decrease shelflife.

Limitations
Gluconic acid is completely metabolised in the body in the same way as a carbohydrate. In the EU, it is included in Part C Group 1 of Regulation 1129/2011, additives permitted at *quantum satis*.

Typical Products
Soft drinks, confectionery and fruit preparations.

Number	Product
E 575	Glucono-delta-lactone (GdL)

Sources
Glucono-delta-lactone is a neutral cyclic ester of gluconic acid, produced with gluconic acid (E 574) by the oxidation by fermentation of glucose or glucose-containing raw materials such as glucose syrup. It is separated from the acid by crystallisation and the material of commerce is a white crystalline powder or granule.

Function in Food
When dissolved in water, GdL hydrolyses to gluconic acid, so it is used as a slow-release acidifier, for curdling milk proteins in cheese and soya proteins in tofu manufacture, and in the ripening process of a wide range of sausages. In bakery, the released gluconic acid serves as a chemical leavening agent by reacting with sodium bicarbonate to give carbon dioxide, and in other products it acts as part of the preservation system. The chelating ability of the acid is particularly useful where iron and copper are present.

Benefits
The principal benefit of GdL is the slow and steady release of the acid in the food product. This lies at the heart of most of the applications. In precipitation of proteins, the steady reduction of pH has a number of benefits in product quality and reduction of production times, and the mild flavour is more compatible with the final product taste than is the case with most organic acids. In some cases it is used to wholly or partially replace bacterial starter cultures. It has a similar role in the processing of meat products, where it may also allow a reduction in the nitrite usage. In bakery products, mixtures of the lactone with sodium bicarbonate release carbon dioxide slowly and continuously, in a comparable way to yeast, but often in a shorter time. It has advantages over some other leavening agents, being less sharply acid and having no bitter or

soapy aftertaste. In gelling mixtures with alginate, the steady decrease in pH improves the final gel texture.

The chelation of iron, copper and other metal ions inhibits oxidation reactions and extends the shelflife of foodstuffs.

Limitations

GdL is completely metabolised in the body in the same way as a carbohydrate. In the EU, GdL is included in Group1 Part C of Regulation 1129/2011, additives permitted at *quantum satis*. JECFA has allocated a group ADI of "not specified" for GdL and the gluconates (E 576–E 578).

GdL must be kept cool and dry as exposure to moisture will initiate the hydrolysis to the acid and lumping of the product.

Typical Products

Cottage cheese, Feta-type cheese, tofu, sausages, refrigerated and frozen prepared doughs.

Number	Product
E 576	Sodium gluconate
E 577	Potassium gluconate
E 578	Calcium gluconate
E 579	Ferrous gluconate

Sources

Sodium gluconate is made by neutralising the mixture resulting from the fermentation of glucose to produce gluconic acid and glucono-delta-lactone with sodium hydroxide.

Potassium and calcium gluconates are made by neutralising the mixture with the appropriate hydroxide or carbonate.

Ferrous gluconate is made either by neutralising gluconic acid with ferrous carbonate or by reacting calcium gluconate with ferrous sulfate.

The sodium, potassium and calcium gluconates are white powders or crystals, which are soluble in water.

Ferrous gluconate is a pale yellowish-grey or green powder, which is soluble in water.

Function in Food

The major uses of sodium gluconate are outside the food industry. Its ability to complex metal ions leads to extensive uses in cleaning products and in the construction industry. In the food industry, its main use derives from its property of covering or reducing the bitterness of other ingredients. Thus, it is used in low-sugar products to cover the bitter or metallic taste of saccharin.

Potassium and calcium gluconates are used mainly for fortification, but the calcium salt is also used as a readily soluble form of calcium, for example in the setting of dessert mixes or the precipitation of proteins. Potassium gluconate is used as a buffer in soft drinks and as a component of salt replacers.

The ferrous salt finds its principal application as a pro-oxidant in the darkening of green olives by an oxidation reaction.

Benefits
The gluconates have a number of advantages over other salts used for fortification. These include high bioavailability, good water solubility and neutral taste.

Limitations
The use of gluconates in fortification of foods is controlled in the EU by Regulation 1925/2006. As additives, the sodium, potassium and calcium salts are included in Group1 Part C of Regulation 1129/2011, additives permitted at *quantum satis*.
Ferrous gluconate is permitted in this Regulation only for use in olives darkened by oxidation to a maximum of 150 mg/kg.

Typical Products
Sodium gluconate in low-calorie chewing gum; Potassium gluconate in baked goods, milk drinks, sport and health drinks and nutritional bars; Calcium gluconate in drinks, desserts and dairy products; Ferrous gluconate in darkened olives.

Number	Product
E 585	Ferrous lactate

Sources
Ferrous lactate is made by reacting ferrous sulfate with calcium or sodium lactate.

Function in Food
Ferrous lactate is used to preserve and darken the colour of olives.

Limitations
Ferrous lactate becomes darker and less soluble in water on exposure to air, so it must be kept, as a raw material, in a tightly sealed opaque container.
In the EU, in Regulation 1129/2011 it is permitted only in olives darkened by oxidation to a maximum of 150 mg/kg.

Number	Product
E 586	4-hexylresorcinol

Sources
4-hexylresorcinol is synthesised from petroleum-based chemicals or glucose. It has a long history of use as a skin disinfectant in soaps and handwashes, and in antiseptic throat sprays and lozenges.

Function in Food
Black spots form on the shells of crustaceans after harvesting because of the action of oxygen on colourless polyphenols, forming first coloured quinones and eventually dark melanins. 4-hexylresorcinol is used as an antioxidant to inhibit this process.

Benefits
4-hexylresorcinol is used in place of sulfites in the prevention of oxidative browning.

Limitations
In the EU, in Regulation 1129/2011, fresh, frozen or deep frozen crustacean meat is permitted to contain residues of 4 hexyl resorcinol up to a maximum of 2 mg/kg.

Number	Product
E 620	Glutamic acid
E 621	Monosodium glutamate
E 622	Monopotassium glutamate
E 623	Calcium diglutamate
E 624	Monoammonium glutamate
E 625	Magnesium diglutamate

Sources
Glutamic acid is an amino acid abundant in nature, either alone or as a component of proteins. As an individual amino acid, it is present in tomatoes and seaweed. The glutamates are made by fermentation, usually starting from starch or molasses. The product of fermentation is separated by filtration, dissolved and neutralised with the appropriate alkaline salt such as sodium hydroxide.
All the forms of glutamic acid and its salts are commonly referred to in the industry as glutamate.

Function in Food
Glutamate is used to develop and enhance the flavour of, mainly, savoury products. It also has its own characteristic flavour, which is considered by some people to be a fifth basic taste, "umami", in addition to the original four of sweet, salt, sour and bitter.
The ammonium and potassium salts are also used in formulations to provide salty taste in low-salt formulations.

Benefits
Glutamate works in a wide variety of dishes, strengthening, developing and rounding savoury flavour.
Of its salts, which are all used in the food industry, monosodium glutamate (MSG) is the one used to the greatest extent in the food industry.

Limitations
Within the EU, the glutamates are included in Regulation 1129/2011 where they are permitted to a maximum of 10g/kg individually or in combination in most food categories with the exception of the salt substitutes and seasonings categories where they are permitted at *quantum satis*. Some exceptional foods, Parmesan cheese for example, naturally contain glutamate at levels higher than 10 g/kg.

The taste of MSG has a self-limiting characteristic. Once the correct amount has been used, any additional quantity contributes little, if anything to the flavour of the product.

Typical Products
Soups, sauces, prepared meals and sausages.

Number	Product
E 626	Guanylic acid
E 627	Disodium guanylate
E 628	Dipotassium guanylate
E 629	Calcium guanylate

Sources
The only guanylate used to any significant extent is the disodium salt, which is produced by the following methods:

1) Enzymatic hydrolysis of yeast ribonucleic acid (RNA), followed by removal of other nucleotides and neutralisation by sodium hydroxide.
2) Fermentation of a sugar source into guanosine, followed by phosphorylation and neutralisation steps.
3) Fermentation of a sugar source into guanylic acid, followed by neutralisation.

Function in Food
Guanylate is the substance responsible for the flavour-enhancement function of shiitake mushrooms. Disodium guanylate is commonly added to food products as a 1:1 mixture with disodium inosinate.

Benefits
The flavour enhancement function of disodium guanylate is synergistically increased when combined with a glutamate source such as monosodium glutamate. The flavour of disodium guanylate is 2.4–3 times stronger than that of disodium inosinate. In addition to enhancing savoury flavour, disodium guanylate also smoothes acidity and saltiness, suppresses bitterness and metallic notes, and masks off-flavours of protein hydrolysates and yeast extracts.

Limitations
In the EU the guanylates are included in Regulation 1129/2011 where they are permitted to a maximum of 500 mg/kg individually or in combination with other guanylates and inosinates, with the exception of the salt substitutes and seasonings categories where they are permitted at *quantum satis*.

Typical Products
Soups, sauces, processed meat, poultry and seafood.

Number	Product
E 630	Inosinic acid
E 631	Disodium inosinate
E 632	Dipotassium inosinate
E 633	Calcium inosinate

Sources
Of the inosinates, only the disodium salt is used to any significant extent. It is produced by the following methods:

1) Enzymatic hydrolysis of yeast ribonucleic acid (RNA), followed by removal of other nucleotides and neutralisation by sodium hydroxide.
2) Fermentation of a sugar source into inosine, followed by phosphorylation and neutralisation steps.
3) Fermentation of a sugar source into inosinic acid, followed by neutralisation.

Function in Food
Inosinate is naturally present in protein foods. It is used to enhance the flavour of red meat, poultry and seafood. Simply adding more meat may not be as effective in increasing the flavour as the concentration of flavour substances is relatively low and they are released only slowly from the tissue of the meat by chewing. Disodium inosinate is commonly added to food products by itself or as a 1:1 mixture with disodium guanylate.

Benefits
The flavour enhancement function of disodium guanylate is synergistically increased when combined with a glutamate source such as monosodium glutamate. Disodium inosinate is a less effective enhancer than the disodium 5'ribonucleotides and disodium guanylate, being 0.33–0.44 times as strong as the guanylate. In addition to enhancing savoury flavour, disodium guanylate also smoothes acidity and saltiness, suppresses bitterness and metallic notes, and masks off-flavours of protein hydrolysates and yeast extracts.

Limitations
In the EU, the inosinates are included in Regulation 1129/2011 where they are permitted to a maximum of 500 mg/kg individually or in combination with other guanylates and inosinates, with the exception of the salt substitutes and seasonings categories where they are permitted at *quantum satis*.

Typical Products
Soups, sauces, processed meat and seafood.

Number	Product
E 634	Calcium 5′ribonucleotides

Sources
Calcium 5′ribonucleotides are obtained by reacting sodium 5′ribonucleotides and calcium chloride.

Function in Food
Calcium 5′ribonucleotides comprise approximately a 1:1 mixture of calcium 5′-guanylate and calcium 5′-inosinate. They are added to food products as flavour enhancers.

Benefits
The flavour enhancement function of calcium ribonucleotides is synergistically heightened when combined with glutamate sources such as monosodium glutamate. The enhancing ability of calcium 5′-ribonucleotides is equal to that of disodium 5′-ribonucleotides (E 635). They are far less soluble in water than disodium 5′-ribonucleotides, and are therefore less susceptible to phosphatase enzymes, which remove the phosphate moiety from the nucleotide molecule. Other functions of the additive include smoothing sharp acidity and saltiness, suppression of bitterness and metallic notes and masking off-flavours of protein hydrolysates and yeast extracts.

Limitations
In the EU calcium 5′ribonucleotides are included in Regulation 1129/2011 where they are permitted to a maximum of 500 mg/kg individually or in combination with other guanylates or inosinates in most food categories, with the exception of the salt substitutes and seasonings categories where they are permitted at *quantum satis*.
The usage of calcium 5′ribonucleotides is very limited in Europe.

Typical Products
Processed seafood, processed meat and poultry and fermented soya-bean soup (miso) bases.

Name	Products
E 635	Disodium 5'ribonucleotides

Sources
Disodium 5'ribonucleotides are produced by the following methods:

1) Enzymatic hydrolysis of yeast ribonucleic acid (RNA), followed by removal of other nucleotides and neutralisation by sodium hydroxide.
2) Fermentation of a sugar source into guanosine and inosine, followed by phosphorylation and neutralisation steps.
3) Fermentation of a sugar source into disodium 5'-guanylate and disodium 5'-inosinate, followed by a mixing process.

Function in Food
Disodium 5'ribonucleotides comprise approximately a 1:1 mixture of disodium 5'-guanylate and disodium 5'-inosinate. They are the most commonly used form of nucleotide flavour enhancers.

Benefits
The flavour enhancement function of disodium 5'-ribonucleotides is synergistically heightened when combined with glutamate sources such as monosodium glutamate. Typically, they are combined with MSG at a ratio of between 2:98 and 10:90. The relative intensity of disodium 5'-ribonucleotides is 1.65–2 times that of disodium inosinate. Other functions of the additive include smoothing sharp acidity and saltiness, suppression of bitterness and metallic notes and masking of off-flavours of protein hydrolysates and yeast extracts.

Limitations
In the EU sodium 5'ribonucleotides are included in Regulation 1129/2011 where they are permitted to a maximum of 500 mg/kg individually or in combination with other guanylates and inosinates in most food categories, with the exception of the salt substitutes and seasonings categories where they are permitted at *quantum satis*.

Typical Products
Soups, sauces, snack seasonings, processed meat and poultry, processed seafood cheese products, tomato-based products and other common processed foods.

Number	Product
E 640	Glycine and its sodium salt

Sources
Glycine is a naturally occurring amino acid that is a part of most proteins. The material of commerce is produced synthetically.

Function in Food
Glycine is used in the food industry as a preservative, an antioxidant and as a browning and seasoning agent.

Benefits
Glycine has a naturally sweet taste and is used alone or as an enhancer of savoury flavours. It can also enhance the taste of saccharin and mask the bitter aftertaste of intense sweeteners.
The Maillard reaction between sugars and amino acids is the reaction that produces browning and flavour development in roasted and baked products from roast meat to cakes. The presence of free amino acids increases the rate of this reaction and glycine is used for this purpose.
Glycine is also used to chelate metal ions that would otherwise catalyse auto-oxidation reactions. It has an inhibitory effect on bacteria but little effect on moulds and yeast.

Limitations
Glycine and its sodium salt are in listed in Group 1 Part C of Regulation 1129/2011 in the EU, additives permitted at *quantum satis*.

Typical Products
Meat products, dietetic foods and salt replacers.

Number	Product
E 650	Zinc acetate

Sources
Zinc acetate is made by the action of acetic acid (E 260) on zinc carbonate or zinc metal. It is available as both an anhydrous compound and as the dihydrate. Zinc is an essential trace element for both animals and humans, and is naturally present in meat and in seafood. It is an integral component of many metalloenzymes.

Function in Food
Zinc acetate is added to chewing gum as a flavour enhancer. It is added to provide an astringent taste, and particularly to intensify the taste of bitterness from ingredients such as coffee or grapefruit.

Limitations
Zinc acetate is permitted in the EU in Regulation 1129/2011 only in chewing gum to a maximum of 1000 mg/kg.

Number	Product
E 900	Dimethyl polysiloxane
	Silicone
	Silicone oil
	Dimethyl silicone

Sources
Dimethyl polysiloxane is made from silica and oil-derived chemicals.

Function in Food
Dimethyl polysiloxane is a surfactant. In foams it occupies the air/water interface and can act in two ways; at low concentrations it is a foam stabiliser and at higher concentrations it causes foams to collapse.

Benefits
Dimethyl polysiloxane is mainly used as an antifoaming agent but is also used to stop hot liquids sticking to equipment. It is useful in products like jam, preventing frothing when boiling, and in carbonated drinks that tend to froth when being filled into bottles or cans.

Limitations
It is permitted in the EU in Regulation 1129/2011 in a range of products: soups, jam, batters, fruit spreads, canned and bottled fruit, confectionery, soft drinks, cider and frying oils to a maximum of 10 mg/kg. It is also permitted in chewing gum to a maximum of 100 mg/kg.
It has been allocated an ADI of 1.5 mg/kg body weight by EFSA.

Typical Products
Soft drinks, jams and catering cooking oil.

Number	Product
E 901	Beeswax

Sources
Beeswax is purified from naturally produced honeycomb.

Function in Food
Beeswax is used to provide a surface finish to stop sticky items sticking together and also to impart a shine to the surface of products. It can also be used on the surface of fresh fruit to reduce drying during storage.

Benefits
Beeswax is a soft wax that allows it to be easily applied to soft sticky items such as gums. To panned goods it gives an instant shine but, being a soft wax, this does not last a long time. It is used in blends with other, harder, waxes to make them easier to apply and to give good adhesion to the piece. Beeswax has the best taste of all the waxes.

Limitations
Within the EU beeswax is permitted in Regulation 1129/2011 at *quantum satis* as a glazing agent in a range of snacks, confectionery and chocolate-coated bakery items, dietary supplements and on the surface of some fresh fruit and as a carrier for flavours.

Typical Products
Small confectionery items such as jellies and gums, tablets.

Number	Product
E 902	Candelilla wax

Sources
Candelilla wax is a natural wax extracted from the leaves of the candelilla plant, *Euphorbia antisyphilitica* Zucc.

Function in Food
Candelilla was the original wax used to plasticise chewing gum. It is used to impart a shine to small pieces of confectionery and to coat the surface of fruit to reduce drying during storage.

Benefits
Candelilla wax is softer and thus easier to apply than carnauba wax. Because it blends with oils, it aids the retention of flavours in gum. Candelilla wax provides the best moisture barrier of the natural waxes.

Limitations
Within the EU, Candelilla wax is permitted in Regulation 1129/2011 at *quantum satis* as a glazing agent in a range of snacks, confectionery and bakery items, dietary supplements and on the surface of some fresh fruit.

Typical Products
Chewing gum, small pieces of confectionery.

Number	Product
E 903	Carnauba wax

Sources
Carnauba wax is a natural material extracted from the fronds of the Brazilian wax palm, *Copernicia prunifera* (Mill.).

Function in Food
Carnauba wax is used to provide a shine to the surface of small sweets. It can also be used to coat the surface of fruit to reduce drying during storage.

Benefits
Carnauba wax is the hardest known wax with a melting point of 78–88 °C. Hard waxes are better for keeping a shine on products throughout the shelflife, but may need to be blended with softer waxes to improve the ease of application.

Limitations
Within the EU, carnauba wax is permitted in Regulation 1129/2011 at *quantum satis* as a glazing agent in a range of snacks, confectionery and bakery items, dietary supplements and on the surface of some fresh fruit.

Carnauba wax has been allocated an ADI of 0–7 mg/kg body weight by EFSA. Carnauba wax works well where pressure can be applied to produce the shine. This makes it inappropriate for delicate products.

Typical Products
Small pieces of chocolate confectionery, gums, jellies and chewing gum.

Number	Product
E 904	Shellac

Sources
Shellac is purified and refined from a resinous secretion of an Indian scale insect, *Laccifer lacca*. It has been used in foods for hundreds of years.

Function in Food
Shellac has a number of uses. It is used to provide a polished surface on confectionery products, to prevent sticky items from sticking together and to reduce moisture loss in stored fruit.

Benefits
Shellac is insoluble but dispersible in water and provides a high gloss without the need for pressure. It can be used in combination with other glazing agents.

Limitations
In the EU, in Regulation 1129/2011, shellac is permitted to be used at *quantum satis* for the surface treatment of fresh fruit and as a glazing agent in confectionery, snacks, chocolate-coated bakery items and food supplements.

Typical Products
Small, bagged confectionery items.

Number	Product
E 905	Microcrystalline wax

Sources
Microcrystalline wax is a hydrocarbon wax isolated in the petroleum industry and subsequently purified.

Function in Food
Microcrystalline wax is a chemically inert wax that is used as a lubricant for chewing gum and to stop sticky products sticking together.

Limitations
Microcrystalline wax is only permitted in the EU in Regulation 1129/2011 for the surface treatment of chewing gum, confectionery decorations and certain fruit where the skin is inedible.

Typical Products
Tropical fruit.

Number   Product
E 907   Hydrogenated poly-1-decene

Sources
Hydrogenated poly-1-decene is prepared by the hydrogenation of mixtures of trimers, tetramers, pentamers and hexamers of 1-decene. Pure 1-decene itself is made from ethylene. Minor amounts of molecules with carbon numbers less than 30 may be present in the mixture. It is insoluble in water, slightly soluble in ethanol and soluble in toluene.

Function in Food
Hydrogenated poly-1-decene is used as a glazing agent on dried fruit and sugar confectionery to reduce loss of moisture during storage.

Benefits
Hydrogenated poly-1-decene is thermally and microbiologically very stable, with very low volatility since the boiling range starts only at 320 °C.

Limitations
It has been allocated an ADI by JEFCA of 0–6 mg/kg body weight. Within Regulation 1129/2011 in the EU it is permitted only for glazing dried fruit and sugar confectionery at levels of up to 2 g/kg.

Typical Products
Dried fruit.

Number   Product
E 912   Montan acid esters
                                          Montan wax

Sources
Montan wax is a hard, brown wax obtained by solvent extraction of a fossilised vegetable wax. It is refined, oxidised with chromic acid and esterified to give a pale yellow wax.

Function in Food
Montan wax is used to provide a protective layer on the skin of fresh fruit to reduce moisture loss during storage.

Limitations
Montan wax is only applied to fruit of which the skin is not eaten. The wax has an unpleasant taste. In the EU, in Regulation 1129/2011, it is only permitted, at *quantum satis*, for use on the skins of fresh citrus and certain tropical fruit.

Number
E 914

Product
Oxidised polyethylene wax

Sources
Oxidised polyethylene wax is prepared by the mild air oxidation of polyethylene.

Function in Food
Oxidised polyethylene wax is used to provide a protective coating to fresh fruit to reduce moisture loss during storage. During the washing process, fruit tends to lose some of its natural wax and glazing agents, such as oxidised polyethylene wax, are used to replace it. The wax is sprayed on as a very small droplet size emulsion.

Limitations
Within the EU, in Regulation 1129/2011, oxidised polyethylene wax is permitted at *quantum satis* only for the surface treatment of fresh citrus and certain tropical fruit.

Number
E 920

Product
L-cysteine hydrochloride

Sources
L-cysteine is a high sulfur-containing aminoacid synthesised by the liver in humans. Industrially it is made by extraction or fermentation of human hair, chicken or duck feathers, or pig bristle.

Function in Food
L-cysteine is used as a flour treatment agent. It is also used in the production of flavourings based on the Maillard reaction.

Benefits
L-cysteine is used to help the gluten relax in doughs that are heavily manipulated, such as pizza bases.

Limitations
In the EU, in Regulation 1129/2011, L-cysteine is only permitted, at *quantum satis*, in flour.

Typical Products
Pizza bases and hamburger buns.

Number	Product
E 927b	Carbamide

Sources
Carbamide, also known as urea, $(CO(NH_2)_2)$ is a white crystalline material made by reaction of carbon dioxide with ammonia followed by purification. It is produced in large volumes for agricultural and pharmaceutical use.

Function in Food
When foods containing fermentable carbohydrate, such as starch or sugars are consumed, bacteria in the mouth use some of the carbohydrate and convert it to acid. When the saliva becomes acid, the teeth lose enamel and cavities begin to form. Chewing gum is often used after meals to generate a flow of saliva and, in so doing, to decrease the acidity of the saliva and thus the rate of development of dental caries. Carbamide is included in the chewing gum to assist in this process of decreasing the acidity of saliva.

Limitations
Carbamide is permitted in the EU, in Regulation 1129/2011, only in chewing gum with no added sugar to a maximum level of 30 g/kg.

Number	Product
E 938	Argon

Sources
Normal atmospheric air contains approximately 0.9% argon. To produce argon, air is filtered, dried and compressed. The compressed air is then allowed to expand, which cools it down so that it liquefies. The liquid air is separated into its components in distillation columns. The argon can be withdrawn as a gas or a cryogenic liquid. Purity levels are typically better than 99.995% after purification.

Function in Food
Argon is used to replace air within food packages in order to extend shelflife by reducing chemical and microbiological activity.

Benefits
In the modified-atmosphere packaging system, argon may be used as an alternative to nitrogen, where a physically inert atmosphere is required. It is denser and more soluble than nitrogen so that, in some situations, less is required than if nitrogen were used.

Limitations
Argon has been allocated an ADI of "not specified" by EFSA. In the EU Argon is included in Part C Group I of Regulation 1129/2011, additives permitted at *quantum satis*. Because argon is a very small proportion of air, it is relatively rare so that its cost, in comparison to other packaging gases is high.

Being denser than air, argon tends to sink so that consideration must be given to safety when it is used in confined or low-lying working environments.

Typical Products
Snack foods, cooked meats, pizzas, recipe dishes and wine.

Number	Product
E 939	Helium

Sources
The major source of helium is from natural gas wells. The gas is obtained by liquefaction and separation.

Function in Food
There are no known commercial applications for helium in food processing or packaging.

Limitations
Helium has been allocated an ADI of "not specified" by EFSA and is included, in the EU, in Part C Group I of Regulation 1129/2011, additives permitted at *quantum satis*.

Number	Product
E 941	Nitrogen

Sources
Nitrogen is the major constituent (78.1%) of air. To prepare pure nitrogen, air is filtered, dried, compressed and cooled by expansion to form a liquid, which is then fractionated in a distillation column. The nitrogen can be provided as a liquid or a gas and purity levels typically exceed 99.995%

Function in Food
Nitrogen is used to extend shelflife in foods by replacing air in food packages, whipping foams, creams and mousses. It is also used to sparge liquids, particularly oils, to remove dissolved oxygen.
Nitrogen is also used widely for food freezing and chilling,

Benefits
Nitrogen is used to extend shelflife by reducing both chemical and microbiological causes of product deterioration.
The oxygen in air can contribute to deterioration in stored foodstuffs through rancidity and enzymic browning, which can be inhibited if the oxygen is replaced by another gas. In addition, many micro-organisms require oxygen in order to grow, so that replacement of the oxygen will inhibit the growth of these aerobic spoilage organisms.

Nitrogen has low solubility in both water and fats and so it is often used in combination with carbon dioxide where the absorption of carbon dioxide could cause packaging collapse.

Limitations
Within the EU, nitrogen is included in Part C Group I of Regulation 1129/2011, additives permitted at *quantum satis*. It has been allocated an ADI of "not specified" by EFSA. Nitrogen is inert and has a small bacteriostatic effect so it is best used where products are initially bacteriologically clean or on its own with low water-content products.

Typical Products
Dried foods, snack food, dried milk and potato powder.

Number   Product
E 942   Nitrous oxide

Sources
Nitrous oxide is most commonly obtained by the thermal decomposition of ammonium nitrate. It may also be obtained by the controlled reduction of nitrates or nitrites or the thermal decomposition of hydroxylamine.

Function in Food
Nitrous oxide is used in dairy products to both provide foaming and to extend shelflife.

Benefits
Nitrous oxide is permitted as a packaging gas that could be used widely, but in practice it is only used in foamed dairy products where it provides both the foaming and the necessary inhibition of oxidative rancidity.

Limitations
Within the EU nitrous oxide is included in Part C Group I of Regulation 1129/2011, additives permitted at *quantum satis*. It has been allocated an ADI of "not specified" by EFSA. Nitrous oxide is a very reactive gas, requiring similar safety measures to oxygen.

Typical Products
Ready-to-serve whipped cream.

Number   Product
E 943a   Butane
E 943b   Iso-butane

Sources
Butane and iso-butane are obtained from liquefied natural gas by fractionation. They are colourless, odourless, flammable gases at normal temperatures and

pressures. They are readily liquefied under pressure at room temperature and are stored and shipped as liquids.

Function in Food
Butane and iso-butane are used as propellant gases for vegetable-oil pan sprays and water-based emulsion sprays.

Limitations
Butane and iso-butane have been allocated an ADI of "not specified" by EFSA. In the EU they are permitted in Regulation 1129/2011 at *quantum satis* only for vegetable-oil pan spray (for professional use only) and water-based emulsion spray.

Number	Product
E 944	Propane

Sources
Propane is a colourless and odourless component of natural gas from which it is obtained by fractionation. It is a gas at normal temperatures and pressures but is readily liquefied under pressure at room temperature and is stored and shipped as a liquid.

Function in Food
Propane is used as a propellant gas.

Limitations
Propane has been allocated an ADI of "not specified" by EFSA. In the EU it is permitted in Regulation 1129/2011 at *quantum satis* only for vegetable-oil pan spray (for professional use only) and water-based emulsion spray.

Number	Product
E 948	Oxygen

Sources
Oxygen comprises 20.9% of air, from which it is extracted by a process of filtration, drying, compression, liquefaction and fractional distillation. It is commercially available as a gas or a liquid. Purity levels are typically better than 99.6%.

Function in Food
Shelflife of food products is normally improved by the exclusion of oxygen but there are some cases where including oxygen in the package can be beneficial. Thus, it is used with packaged fresh red meat where a high oxygen content preserves the redness of the meat. Here, the oxygen content may be as much as 80% with the remainder being carbon dioxide.

Oxygen is also used in a low oxygen/high nitrogen environment to reduce the respiration rate and extend the shelflife of prepared salads. Here, the combination may be 95% nitrogen and 5% oxygen.

It is also used to produce an aerobic atmosphere, with carbon dioxide and nitrogen, for packing fresh white fish.

Benefits

Oxygen is used in the packaging of fresh meat products to preserve the red colour of the meat, maintaining its attractiveness. The packaging gases are chosen to maximise the length of time the meat remains red and to reduce the chance of microbiological spoilage.

The low-oxygen atmosphere of prepared packaged salads reduces the respiration rate and reduces the enzymic browning on the cut surfaces. The low-oxygen environment is maintained by matching the rate of oxygen reduction with a packaging film of similar permeability, to maintain an equilibrium-modified atmosphere.

Oxygen is used in fish packaging as complete removal of oxygen may permit the growth of pathogenic anaerobic organisms such as *Clostridium botulinum*.

Limitations

Oxygen has been allocated an ADI of "not specified" by EFSA. In the EU it is included in Part C Group I of Regulation 1129/20011, additives permitted at *quantum satis*. Oxygen allows the growth of aerobic micro-organisms and suitable hurdles must be in place to ensure food is safe to eat over its entire shelflife.

It is a highly reactive gas and when a high oxygen atmosphere is used in packaging, care must be taken to ensure the machinery is oxygen compatible and that necessary procedures are in place to permit its safe use.

Typical Products

Fresh meat, cut and prepared salads, white fish and seafood.

Number	Product
E 949	Hydrogen

Sources

Hydrogen is produced by catalytically reforming natural gas or other hydrocarbon fuels with superheated steam at elevated temperatures. It can also be produced by electrolytic decomposition of alkalised water, and by the cracking of gaseous ammonia.

Function in Food

The principle use of hydrogen in the food industry is in the processing of fats and oils. Used with a catalyst, such as nickel, hydrogen converts double bonds in unsaturated oils to single bonds creating stable, saturated fats. Hydrogen has also been considered as an innovative component of gas mixtures for modified-atmosphere packaging applications.

Limitations
Hydrogen has been allocated an ADI of "not specified" by EFSA. In the EU it is permitted in Part C Group I of Regulation 1129/2011, additives permitted at *quantum satis*.

Number	Product
E 950	Acesulfame K

Sources
Acesulfame K is the potassium salt of 6-methyl-1,2,3-oxathiazine-4(3H)-one-2,2-dioxide. Several synthetic routes have been described.

Function in Food
Acesulfame K is a noncariogenic, nonlaxative, intense sweetener used in a wide range of foods, including foods for diabetics. Since it is not metabolised by the human body, it passes through the digestive system unchanged and is, therefore, completely "calorie-free". It is approximately 130–200 times as sweet as sucrose and, in addition, can act as a flavour enhancer.

Benefits
Acesulfame K provides a clean sweetness with a fast onset. With other intense sweeteners, acesulfame K shows synergistic effects, which leads to a more sugar-like taste and additional sweetness enhancement. In particular, blends of acesulfame K and aspartame have a high synergy. Other synergistic taste enhancements have been demonstrated in blends of acesulfame K and alitame, cyclamate, neohesperidine DC and sucralose.
Acesulfame K is stable to a wide range of processing conditions, tolerating pH levels from 3 to 9 and temperatures up to 200 °C. It is highly soluble in water.

Limitations
Within the EU, acesulfame K is permitted in Regulation 1129/2011 at *quantum satis* in table-top sweeteners and in a range of energy-reduced or no-added-sugar products with individual maxima in each case.
When used in high concentrations above normal use levels, acesulfame K may have a slight aftertaste. The ADI has been set at 0–15 mg/kg by JECFA but 0–9 mg/kg body weight by the SCF. Typical usage levels are 100–300 mg/l in beverages and up to 2000 mg/kg in confectionery and baked goods.

Typical Products
Beverages, yogurts, ice cream and other dairy products, desserts, confectionery, chewing gum and table-top sweeteners.

Number   Product
E 951   Aspartame

Sources
Aspartame is the methyl ester of a dipeptide composed of the amino acids
L-aspartic acid and L-phenylalanine. The production of aspartame normally
starts from L-phenylalanine or L-phenylalanine methyl ester and L-aspartic
acid. The synthesis follows the common routes of peptide synthesis as the
L-configuration of the amino acids has to be retained. An alternative to
chemical peptide synthesis is enzymic formation of the peptide bond; both
processes are used commercially.

Function in Food
Aspartame is a nutritive, noncariogenic intense sweetener with sweetness
potency approximately 200 times that of sucrose. In addition, it can act as a
flavour enhancer, most noticeably with fruit flavours. Aspartame is digested as
it is a dipeptide and it has the same calorific value as sugar (4 kcal/g)
but because considerably less is used in a product, it acts to reduce the
calorie content of the food. Aspartame is nonlaxative and suitable for diabetics.

Benefits
Aspartame provides a clean sweet taste, which can be slightly delayed and
longer lasting than that of sugar. It can be combined with other intense
sweeteners and with carbohydrate sweeteners. It is synergistic with acesulfame
K and the combination has a more sugar-like taste (masking the lingering
sweetness of aspartame). Typical usage levels are between 600 mg/l in beverages
and up to 5500 mg/kg in chewing gums with no added sugar.

Limitations
Although aspartame is relatively stable in dry form, pH, temperature and time
are very important factors affecting its stability in solution. The maximum
stability of aspartame is obtained at pH4.2–4.3. The hydrolysis between pH3
and pH5 can be limited under controlled temperature. Below pH3 and above
pH5, aspartame decreases rapidly even under ambient storage conditions.
Therefore, aspartame is not very suitable for applications such as baked goods,
since the manufacturing process involves exposure both to high pH levels and
to high temperatures.
JEFCA has allocated an ADI for aspartame at 0–40 mg/kg. Within the EU,
aspartame is permitted in Regulation 1129/2011 at *quantum satis* in table-top
sweeteners and in a range of products with individual maxima.
Persons suffering from the genetic condition phenylketonuria (PKU) must be
aware of the presence of phenylalanine in their diet. In the EU, Regulation
1333/2008 requires that products containing this sweetener must be labelled
with a statement saying "contains a source of phenylalanine".

Typical Products
Low-energy drinks and table-top sweeteners

Number	Product
E 952	Cyclamic acid and its salts

Sources
Cyclamic acid is a synthetic sweetener that is manufactured by the sulfonation of cyclohexamine. It is then converted to the sodium or calcium salts that are the commercial forms of the sweetener.

Function in Food
Cyclamate is an intense low-calorie sweetener. It is generally considered to be about 30 times as sweet as sucrose.

Benefits
The sweet taste profile of cyclamate builds to a maximum more slowly than that of sucrose, but it also lingers for a longer time. It functions very effectively in combination with other sweeteners – in particular with saccharin, where that sweetener's bitter/metallic aftertaste is masked by cyclamate. When used in combination with saccharin, the normal ratio is 10:1 cyclamate:saccharin. This blend delivers a cost-effective, acceptable, sweet taste profile.
Cyclamate also synergises effectively with aspartame and acesulfame K, with commercial uses being in binary, tertiary or even quaternary blends with these sweeteners. It is particularly compatible with citrus flavours.
Sodium and calcium cyclamate are very soluble in water and solutions are stable at low pH to heat and light.
Cyclamate is noncariogenic and is able to mask bitter tastes. Consequently it is used in oral hygiene products and in liquid pharmaceutical preparations.
Cyclamate is used extensively in Asia. Europe consumes approximately 15% of the world supply.

Limitations
In the EU cyclamate is included in the list of permitted sweeteners in Regulation 1129/2011 where it is permitted in table-top sweeteners *quantum satis* and in energy-reduced or no-added-sugar drinks to a limit of 250 mg/l. It has been allocated an ADI of 0–7 mg/kg body weight by EFSA and 0–11 mg/kg by JEFCA.

Typical Products
Table-top sweeteners and soft drinks.

Number	Product
E 953	Isomalt

Sources
Isomalt is a sugar replacer belonging to the group of polyols. It is a white crystalline disaccharide alcohol produced by enzymic conversion of sucrose

into isomaltulose, followed by hydrogenation into isomalt. It is defined as a mixture of hydrogenated mono- and disaccharides whose principal components are the disaccharides 1-0-α-D-Glucopyranosyl-D-mannitol dihydrate (1,1GPM dihydrate) and 6-0-α-D-Glucopyranosyl-D-sorbitol (1,6-GPS). Depending on the detailed composition of the saccharides and their intended application, a number of commercial variants in a range of particle sizes are available.

Function in Food
Isomalt is a sugar replacer with a sweetness of 50 to 60% that of sugar. Its most important use is in confectionery with a reduced sugar or calorie claim. In these products it provides sweetness, bulk and texture while replacing sugar.

Benefits
The sweet taste of isomalt is similar to that of sugar but with a lower intensity. It is synergistic with most intense sweeteners. Unlike some other polyols, it does not give a cooling effect in the mouth and it dissolves slowly so that sweets last longer. Isomalt is noncariogenic and is not hygroscopic and does not participate in the Maillard reaction. These characteristics can be used to advantage in controlling moisture uptake during shelflife and controlling colour development during cooking.
Only about half of the energy content is actually utilised by the human body. In the EU it is given an energy content of 2.4 kcal/g and in the US 2.0 kcal/g.

Limitations
In the EU isomalt is permitted in Part C Group 4 (polyols) of Regulation 1129/2011. It is permitted at *quantum satis* in table-top sweeteners and in a range of energy-reduced or no-added-sugar products, for example confectionery, dietary products and supplements. It is also permitted for purposes other than sweetening in fresh fish and crustacea.
It has been allocated an ADI of "not specified" by JEFCA.
In the EU, Regulation 1333/2008 requires that if a foodstuff contains more than 10% isomalt, the product must be labelled to the effect that excessive consumption may produce laxative effects.

Typical Products
Hard and soft candies, chewing gum, chocolate products and baked goods.

Number	Product
E 954	Saccharins

Sources
Saccharin is a white crystalline powder synthesised from petroleum-based starting materials. It has been used as a sweetener for over 100 years. It is

available as both the acid and its sodium, potassium and calcium salts, with the most common form being the sodium salt.

Function in Food
The saccharins are intense, noncaloric, noncariogenic sweeteners used to replace sugar in reduced-calorie products.

Benefits
Saccharin is approximately 450 times sweeter than sugar. The acid is insoluble in water but the salts are highly water soluble. The saccharins are stable to a range of processing conditions. They are not metabolised by the human body or by the bacteria that cause dental caries. They are synergistic with other intense sweeteners.

Limitations
The saccharins have a bitter metallic aftertaste. They are approved for use in over 100 countries including the USA and the countries of the EU. In the EU, the saccharins are included in the list of permitted sweeteners in Regulation 1129/2011 where they are permitted at *quantum satis* in table-top sweeteners and in a range of low-sugar or energy-reduced products with individual maxima in each case. EFSA, JEFCA and the FDA have allocated an ADI for sodium saccharin of 0–5 mg/kg body weight.

Typical Products
Table-top sweeteners, energy-reduced drinks, canned fruit and dessert toppings.

Number	Product
E 955	Sucralose

Sources
Sucralose is produced by the selective chlorination of three of the hydroxyl groups of sucrose to produce 1,6-dichloro-1,6 dideoxy–β-D-fructofuranosyl–4-chloro-4-deoxy–α–D-galactopyranoside. The product is purified by crystallisation and dried.

Function in Food
Sucralose is a noncaloric, noncariogenic intense sweetener, approximately 600 times sweeter than sugar. It is used to replace sugar in table-top sweeteners and in a wide range of manufactured foodstuffs.

Benefits
Sucralose is nonhygroscopic. It is synergistic with other intense sweeteners such as Aspartame and Acesulfame K. It also withstands high-temperature

processing and long-term storage even when used in low-pH products such as carbonated drinks.

Limitations
Within the EU, sucralose is permitted in Regulation 1129/2011 at *quantum satis* in table-top sweeteners and in a wide range of products with individual maxima in each case. It has been allocated an ADI of 0–5 mg/kg body weight by the FDA but of 0–15 mg/kg by JEFCA and EFSA.

Typical Products
Table-top sweeteners, soft drinks.

Number	Product
E 957	Thaumatin

Sources
Thaumatin is a protein contained in the fruits of the plant *Thaumatococcus danielli* Bennett, which grows in West Africa. Fruits are harvested and part processed to remove the section known to contain thaumatin. Final processing is based on water extraction, ultrafiltration and freeze drying to produce thaumatin.

Function in Food
Thaumatin is a naturally sweet protein, approximately 2000 times sweeter than sugar, which is used at very low levels, typically 0.5–3 ppm (below the sweetness threshold) for its flavouring properties. It can mask unpleasant tastes and synergise with other flavourings, sweeteners and flavour enhancers to improve the taste and mouthfeel of a wide range of products.

Benefits
Thaumatin can mask bitterness and unpleasant after tastes from soya, intense sweeteners, vitamins, minerals and herbs and can reduce the off-notes arising during manufacture and storage of food and beverages, especially noted in citrus flavours. Thaumatin is synergistic with other ingredients to improve taste and has the advantage of improving mouthfeel in low-fat or low-calorie products. It is not cariogenic and is stable to both heat and pH.

Limitations
In the EU, in Regulation 1129/2011, thaumatin is permitted at *quantum satis* in table-top sweeteners and in a range of energy-reduced or no-added-sugar products and chewing gum with individual maxima in each case. It is also permitted as a flavour enhancer in nonalcoholic drinks. It is considered GRAS in the USA and has an ADI of "not specified" allocated by JEFCA.

Typical Products
Chewing gum, edible ices, nonalcoholic drinks.

Number   Product
E 959   Neohesperidine DC

Sources
Neohesperidine DC is a white crystalline intense sweetener prepared from the waste material from citrus processing.

Function in Food
NHDC is between 1000 and 1800 times sweeter than sugar. The sweet taste develops more slowly and lingers longer than that of sugar. It is used in very small amounts to enhance sweet taste and fruit flavours and to improve mouthfeel.

Benefits
NHDC is stable at high temperatures and has a long ambient shelflife both as a powder and in aqueous solution. NHDC is synergistic with many other sweeteners, both intense sweeteners and polyols.

Limitations
In the EU, NHDC is permitted under Regulation 1129/2011, at *quantum satis* in table-top sweeteners and in a range of energy-reduced or no-added-sugar products with individual maxima in each case. It has a distinctive taste and is unsuitable for use as the sole sweetener in products that do not require a liquorice taste.

Typical Products
Chewing gum, soft drinks, dairy products.

Number   Product
E 960   Steviol glycosides
                                          Stevia

Sources
The plant *Stevia rebaudiana* Bertoni is a leafy bush of the *compositae* family, growing to 40 to 60 cm tall and native to South America, where its leaves have been used for centuries to sweeten foods and beverages. The plant is now cultivated in a number of countries on both sides of the equator including Kenya, Paraguay, Brazil, Indonesia, Thailand, China and Japan. 1 hectare of crop yields approximately 4 tons of leaves, which when processed yield several hundred kilos of steviol glycosides.
The leaves contain 11 different steviol glycosides of differing sweetening power, principally stevioside, rebaudioside A, rebaudioside C and dulcoside A. While each manufacturer has their own particular technique, in general the glycosides are extracted from the leaves using water and the extract purified before pasteurisation and spray drying to give a white to yellowish powder that is soluble in water.

Stevia has a sweetening power of 250–300 times that of sugar, the main contributors being stevioside and rebaudioside A. A number of extracts are commercially available, differing in the ratio of stevioside to rebaudioside A

Function in Food
Stevia is a noncariogenic intense sweetener used in table-top sweeteners and a range of manufactured foodstuffs. It is considered noncaloric in the EU but to contain 0.2 kcal/g in USA.

Benefits
Stevia is nonhygroscopic and is stable as a powder up to 180 °C and in liquids over a range of pH from 3–7. It does not participate in the Maillard reaction. It has a negative heat of solution so provides a cooling effect in the mouth.

Limitations
Within the EU, stevia is permitted in Regulation 1131/2011 at *quantum satis* in table-top sweeteners and in a range of energy-reduced or no-added-sugar products with maximum levels in each case. The maximum levels are, in general, to low to allow stevia to be the sole sweetener in the product. Stevia tends to be blended with sweeteners that are able to moderate the taste profile and improve the "sugar-like" sweetness.
Stevia has been determined to be GRAS for its intended uses in the USA.
EFSA and JECFA have allocated stevia an ADI 0–4 mg/kg body weight expressed as steviol equivalents.

Typical Products
Soft drinks, table-top sweeteners.

Number	Product
E 961	Neotame

Sources
Neotame is a derivative of aspartame prepared by the reaction of aspartame and 3,3-dimethylbutyraldehyde followed by purification, drying and milling.

Function in Food
Neotame is a very intense sweetener, some 7000–13 000 times sweeter than sugar and 30–60 times sweeter than aspartame. It is used as a sweetener and flavour enhancer.

Benefits
Neotame has a clean, sweet taste and can act as a flavour enhancer with certain flavours, particularly mint.
Neotame is moderately heat stable when dry but stability in solution is determined by pH, temperature and time.

It has been determined that very little, if any, phenylalanine is released into the bloodstream as a result of the digestion of neotame so that products containing it do not have to carry a warning notice.

Limitations
EFSA and JECFA have both allocated neotame an ADI of 0–2 mg/kg body weight
Neotame is permitted in Australia and given general approval as a sweetener and flavour enhancer in the USA. In the EU, it is included in Regulation 1129/2011, where it is permitted at *quantum satis* in table-top sweeteners and in a number of energy-reduced or no-added-sugar products to individual maxima in each case.
It degrades slowly in aqueous conditions such as those in soft drinks.

Typical Products
Table-top sweeteners and soft drinks.

Number	Product
E 962	Salt of aspartame and acesulfame

Sources
The salt is prepared by heating an approximately 2:1 ratio (w:w) of aspartame and acesulfame K in solution at acidic pH and allowing crystallisation to occur. It is sparingly soluble in water and slightly soluble in ethanol.

Function in Food
The salt is used as a noncariogenic, intense sweetener, being approximately 350 times sweeter than sugar.

Benefits
The salt is more stable than aspartame alone and acts as a slow-release sweetener in chewing gum.

Limitations
Within the EU, the salt of aspartame and acesulfame K is permitted as a sweetener in Regulation 1129/2011 at *quantum satis* in table-top sweeteners and in a range of energy-reduced or no-added-sugar products with individual maxima in each case. The ADI has been assessed by both JECFA and EFSA and both bodies have concluded that the ADI is covered by the already agreed ADI for the individual components of the salt.
In the EU, Regulation 1333/2008 requires that products containing this sweetener must be labelled with a statement saying "contains a source of phenylalanine".

Typical Products
Table-top sweeteners, beverages, confectionery.

Number	Product
E 965	Maltitol

Sources
Maltitol and maltitol syrups are manufactured by hydrogenation of maltose and maltose/glucose syrup. Maltitol syrups may be described as hydrogenated starch hydrolysate. The syrup has a dry matter content of 75%.

Function in Food
Pure crystalline maltitol is 90% as sweet as sucrose; the sweetening power of maltitol syrups ranges from 60 to 85% of that of sugar.
It is a nutritive sweetener used to replace sucrose and glucose syrups in sugar-free confectionery products such as chocolate, chewing gum and hard-boiled soft and chewy candies. Use of crystalline maltitol provides a crunchy chewing gum coating and helps control texture and flexibility extending storage stability of chewing gum. Maltitol syrup is used as a plasticiser in chewing gum to give a more stable and softer texture.

Benefits
Maltitol is noncariogenic. It is not fully metabolised by the human body and has been allocated an energy content of 2.4 kcal/g in the EU and 2.1 kcal/g (3.0 kcal/g for maltitol syrups) in the US.
It does not participate in the Maillard reaction so does not contribute to browning on cooking.

Limitations
Within the EU maltitol is regulated as a polyol in Part C Group 4 of Regulation 1129/2011 and is permitted at *quantum satis* in table-top sweeteners and in a range of energy-reduced or no-added-sugar confectionery products, chewing gum, desserts and spreads. It is also permitted for purposes other than sweetening in fresh fish and crustacea. As with all polyols excessive use of maltitol can have a laxative effect and in the EU Regulation 1333/2008 requires that products containing more than 10% added maltitol must be labelled to that effect.

Typical Products
Sugar-free confectionery and chewing gum.

Number	Product
E 966	Lactitol

Sources
Lactitol is a sugar alcohol produced by catalytic hydrogenation of lactose. Lactitol exists in three forms – dihydrate, monohydrate and anhydrous. The difference is in the amount of bound water.

Function in Food
Lactitol is a bulk sweetener used in sugar-free, sugar-reduced and low-calorie foods. In surimi, it acts as a cryoprotectant, and prevents denaturation of fish protein during freezing.

Lactitol can also be used as a prebiotic in a range of functional foods such as yogurts and bakery products. In the colon, lactitol can be fermented by beneficial bacteria such as bifidobacteria and *Lactobacillus* species.

Benefits
Lactitol is a disaccharide with similar physical properties to sucrose but with a lower calorific value. In the EU this is determined to be 2.4 kcal/g and in the USA 2.0 kcal/g. It is metabolised independently of insulin and is suitable for diabetics.

It does not participate in the Maillard browning reaction so that it does not contribute to browning during cooking and it is not hygroscopic. It is not fermented by mouth bacteria that cause tooth decay.

Limitations
Lactitol is only 30 to 40% as sweet as sugar. Within the EU it is regulated as a polyol in Group 4, Part C of Regulation 1129/2011 for use at *quantum satis* in table-top sweeteners and in energy-reduced and no-added-sugar confectionery and chewing gum, spreads, breakfast cereals and desserts. It is also permitted for purposes other than sweetening in fresh fish and crustacea. As with all polyols excessive use of lactitol can have a laxative effect and, in the EU, Regulation 1333/2008 requires that products containing more than 10% added lactitol must be labelled to that effect.

Typical Products
Energy-reduced confectionery and chewing gum.

Number	Product
E 967	Xylitol

Sources
Xylitol is produced by the catalytic hydrogenation of xylose that can be obtained from the xylan-rich hemicellulose portion of trees and plants. Xylitol is a natural constituent of many fruits and vegetables at levels of less than 1% and the human body produces 5–15 g of xylitol per day during the metabolism of glucose.

Function in Food
Xylitol is principally used as a nonfermentable bulk sweetener in foods and oral hygiene products. In addition to its use as a sweetener, xylitol is also used as a humectant, as a masking agent for other ingredients, and as an energy source in intravenous products.

Benefits

Xylitol has a similar sweetness profile to that of sucrose, with no discernible aftertaste. In addition, xylitol has a distinct cooling effect in the mouth due to its negative heat of solution (the greatest of all the sweeteners). It is metabolised independently of insulin and consumption does not affect blood glucose levels. It has been determined to have an energy content of 2.4 kcal/g in both the EU and the USA.

It resists fermentation by oral bacteria and inhibits the growth of *streptococcus mutans*, the organism most responsible for dental caries. The ability of xylitol to inhibit the development of new caries has been demonstrated in numerous clinical and field studies.

Limitations

Xylitol is permitted as a sweetener in the EU in Regulation 1129/2011 where it is included in Part C Group 4, polyols. It is permitted at *quantum satis* in table-top sweeteners and in a range of energy-reduced or no-added-sugar confectionery, chewing gum, spreads and desserts. It is also permitted for purposes other than sweetening in fresh fish or crustacea.

The EU Scientific Committee on Food concluded that daily consumption of less than 20 g of polyols was unlikely to cause laxative effects except in sensitive individuals.

However, as with all polyols, the EU Regulation 1333/2008 requires that if a foodstuff contains more than 10% added xylitol it must be labelled "excessive consumption may produce laxative effects".

Typical Products

Chewing gum, no-added-sugar confectionery and toothpaste.

Number	Product
E 968	Erythritol

Sources

Erythritol is obtained by fermentation of glucose by an osmophilic yeast, *Moniliella pollinis* or *Trichosporonoides megachiliensis*. It is soluble in water and slightly soluble in alcohol.

Function in Food

Erythritol is a polyol used as a bulk sweetener to replace sugar. It is used in combination with other polyols and with intense sweeteners. As well as being a sweetener it can act as a flavour enhancer, humectant, bulking agent and sequestrant.

Benefits

Erythritol has 60–70% of the sweetness of sugar and is often blended with intense sweeteners to replicate the sweetness profile of sugar. It is stable to heat and is not hygroscopic. It is not metabolised by the human body and has been allocated an energy content of 0.2 cal/g in USA but is considered calorie-free in Japan and the EU. It is not metabolised by the bacteria that cause tooth decay.

Limitations
Erythritol has a strong cooling effect when it dissolves in the mouth that complements mint flavours.
It is included in the list of polyols permitted in Part C Group 4 of Regulation 1129/2011 permitted for use at *quantum satis* in table-top sweeteners and in a range of energy-reduced or no-added-sugar products, for example confectionery, chewing gum and desserts. It is also permitted for purposes other than sweetening in fresh fish and crustacea.
As with all polyols, EU Regulation 1333/2008 requires that if a foodstuff contains more than 10% added erythritol it must be labelled "excessive consumption may produce laxative effects".

Typical Products
Confectionery, desserts, food supplements

Number	Product
E 999	Quillaia extract

Sources
Quillaia extract is an aqueous extract of the bark of the tree *Quillaia saponaria* Molina. The solution is dried to yield a light brown powder, which is odourless but has an acrid astringent taste. The active constituents are saponins, which are also present in sarsaparilla, liquorice and yucca.

Function in Food
Quillaia extract is used to provide a stable foaming head on soft drinks such as ginger beer and cream soda.

Limitations
In the EU, quillaia extract is permitted in Regulation 1129/2011 in soft drinks and cider, other than cidre bouché, to a maximum level of 200 mg/l.

Typical Products
Ginger beer.

Number	Product
E 1103	Invertase

Sources
Invertase is an enzyme naturally present in human saliva. It is produced commercially by submerged fermentation of yeast, from which it is separated and purified.

Function in Food
Invertase is used to produce invert sugar (a mixture of glucose and fructose) from sucrose within food products.

Benefits
Invertase is used industrially to make invert syrup (golden syrup) from solutions of beet or cane sugar. It is also used in products that are difficult to make with soft centres; the paste for the centre can be made with a firm texture but with the addition of invertase so that it softens after the assembly of the sweet but before consumption. The optimum pH is between 4.5 and 5.5 and the enzyme can work in liquid phases containing as much as 75% sucrose.

Limitations
Invertase has been allocated an ADI of "not specified" by EFSA and, in the EU, is included in Group I of Part C of Regulation 1129/2011, additives permitted at *quantum satis*.

Typical Products
Confectionery products with soft or liquid centres.

Number	Product
E 1105	Lysozyme

Sources
Lysozyme is an enzyme extracted and purified from hen egg albumen.

Function in Food
It is used to inhibit growth of the bacteria in hard cheese that cause "late blowing".

Benefits
Lysozyme can be used in cheese in place of nitrate.

Limitations
Lysozyme has been allocated an ADI of "not specified" by EFSA and, in the EU, is only permitted in Regulation 1129/2011, albeit at *quantum satis*, in ripened cheese and cheese products.

Number	Product
E 1200	Polydextrose

Sources
Polydextrose is prepared by a vacuum melt process involving polycondensation of glucose in the presence of small amounts of sorbitol and an acid. The final

product of this reaction is a weakly acidic water soluble polymer that contains minor amounts of bound sorbitol and acid. The polymer is then subjected to clean up procedures to produce several grades of polydextrose.

Function in Food
Polydextrose is a low-calorie bulking agent that is used as a partial replacement of sugars and/or fats whilst maintaining texture and mouthfeel. In addition, it has two useful technological functions; it can be used to stabilise foods by preventing sugar and polyol crystallisation, *e.g.* in hard candies and it can function as an humectant and retard the loss of moisture in baked goods, which helps protect against staling.

Benefits
Polydextrose is not sweet and can be used for both sweet and savoury products. It is only partially metabolised by bacteria in the large intestine and has been ascribed a calorific value of 1 kcal/g. Because it is digested in a similar way to dietary fibre and its metabolism does not involve insulin, polydextrose can be used in products designed for diabetic and low-glycaemic diets. Polydextrose is not fermented by mouth bacteria and will not promote dental caries.
Polydextrose is also used in conjunction with other materials as a film and tablet-coating agent and solutions are used in pharmaceutical preparations as binders in wet granulation processes.

Limitations
Polydextrose has been allocated an ADI of "not specified" by EFSA and, in the EU is included in Part C Group I of Regulation 1129/2011, additives permitted at *quantum satis*.

Typical Products
No-added-sugar, energy-reduced and dietetic confectionery, frozen dairy desserts, baked goods, surimi and beverages.

Number	Product
E 1201	Polyvinylpyrrolidone

Sources
Polyvinylpyrollidone is made by a multistage synthesis from butan-1,3-diol followed by purification.

Function in Food
Polyvinylpyrollidone is used to help tablets break up in water.

Benefits
Polyvinylpyrollidone is water soluble and it is used in tablet coatings to increase the penetration of water into the tablet.

Limitations
Polyvinylpyrrolidone has been allocated an ADI of "not specified" by EFSA. Within the EU it is permitted at *quantum satis* only in table-top sweeteners and food supplements in tablets or coated tablets under Regulation 1129/2011.

Number   Product
E 1202   Polyvinylpolypyrrolidone

Sources
Polyvinylpolypyrrolidone is a white powder made by crosslinking polyvinylpyrollidone followed by purification.

Function in Food
Polyvinylpolypyrrolidone is used to help tablets break up in water. It is also used as a processing aid in the treatment of wine because it binds strongly to tannins.

Benefits
Polyvinylpolypyrollidone is water absorbent and swells in water, but is not water soluble. When included in a tablet this swelling in water is used as a means of making the tablet break up.

Limitations
Polyvinylpolypyrrolidone has been allocated an ADI of "not specified" by EFSA. Within the EU it is permitted at *quantum satis* only in table-top sweeteners and food supplements in tablets or coated tablets under Regulation 1129/2011.

Number   Product
E 1203   Polyvinyl alcohol

Sources
Polyvinyl alcohol is made by the hydrolysis of polyvinyl acetate to remove acetate groups. While polyvinyl acetate is made by the polymerisation of vinyl acetate, vinyl alcohol exists almost exclusively as the tautomeric form, acetaldehyde, so polyvinyl alcohol cannot be produced directly by polymerisation.

Function in Food
PVA is used as a film-coating agent in food supplements.

Limitations
In the EU, PVA is permitted in Regulation 1129/2011 only in supplements in capsule and tablet form up to a maximum of 18 000 mg/kg.

Number	Product
E 1204	Pullulan

Sources
Pululan is a polysaccharide gum consisting mainly of 1,6-α-linked maltotriose units, produced by fermentation of a hydrolysed food grade starch by the fungal species *Aureobasidium pullulans* followed by filtration, sterilisation and purification. Pullulan is a water-soluble white powder that has been in use in Japan for upwards of 20 years but is newly permitted in the USA and the EU.

Function in Food
Pullulan is used as a glazing agent, a film former and thickener.

Benefit
Pullulan forms transparent, water soluble, fat resistant and flavourless films. It is used as an alternative to gelatine for making capsules.
It is used as a coating to slow tablet deterioration and as a binding agent for tablets.
Pullulan solutions are stable to pH3–8. They are viscous but do not form gels.

Limitations
In the EU, pullulan is permitted in Regulation 1129/2011 at *quantum satis* in breath-freshening microsweets in the form of films and in food supplements in capsule and tablet form. JEFCA has allocated pullulan an ADI of "not specified". In the USA it is considered GRAS.

Typical Products
Mint microsweets.

Number	Product
E 1205	Basic methacrylate copolymer

Sources
Basic methacrylate copolymer is manufactured by controlled polymerisation of the monomers methyl methacrylate, butyl methacrylate and dimethyl-aminoethyl methacrylate.

Function in Food
Basic methacrylate copolymer is used as a glazing or coating agent for solid food supplements in order to provide moisture protection or taste masking of nutrients together with fast release of the nutrient in the stomach.

Benefits
The copolymer is not absorbed by the gastrointestinal tract to any significant extent.

Limitations
In the EU, it is permitted in Regulation 1129/2011 where it is limited for use at
100 000 mg/kg in food supplements in solid form.

Number   Product
E 1404   Oxidised starch

Sources
Native starches are oxidised by treating an aqueous starch suspension with
sodium hypochlorite.

Function in Food
Oxidised starches provide soft gels, which exhibit greater stability in high-sugar
systems and greater resistance to shrinkage. They are also used to improve
adhesion properties of batters.

Benefits
Oxidised starches can be used at higher dosage rates than their parent native
starches, thus increasing the range of textures achievable in gum confectionery.
A further benefit is increased and improved shelflife. Oxidised starches, used in
adhesion batters, can improve the visual and eating quality of battered foods,
owing to a more consistent coverage of the substrate.

Limitations
Within the EU, oxidised starch is included in Part C Group 1 of Regulation
1129/2011, additives permitted at *quantum satis*, except in foods for young
children where they are permitted to a maximum of 50 000 mg/kg.

Typical Products
Gum confectionery, dairy products, batter and bread coatings for meat, fish
and vegetables.

Number   Product
E 1410   Monostarch phosphate

Sources
Native starches are phosphorylated to produce these modified starches (also
referred to as stabilised starches) where only one starch hydroxyl group is
involved in the starch-phosphate linkage. Typical reagents include orthophosphoric acid, sodium or potassium orthophosphate, or sodium tripolyphosphate.

Function in Food
These modified starches are used as freeze–thaw-stable thickeners for simple
processes. For greater process tolerance, crosslinked starches are required.
Starch phosphates also exhibit good emulsifying properties. Pregelatinised

starch phosphates are used as thickeners in dry-mix puddings and as binders in bakery products.

Benefits
Monostarch phosphates improve the product quality and shelflife stability of frozen foods. Product quality is also improved in salad dressings when they are used as emulsion stabilisers. Their incorporation into baked goods can improve moisture retention, which enhances eating quality and extends shelflife.

Limitations
Monostarch phosphates are permitted in the EU in Group 1 Part C of Regulation 1129/2011, additives permitted at *quantum satis* except in foods for young children where they are permitted to a maximum of 50 000 mg/kg. Monostarch phosphates are less frequently used as thickening agents than distarch phosphates.

Typical Products
Pie fillings, salad dressing and frozen gravies.

Number	Product
E 1412	Distarch phosphate

Sources
Native starches are crosslinked by reacting an aqueous starch slurry with reagents such as phosphorus oxychloride or sodium trimetaphosphate to make distarch phosphates. A range of starches is available with different levels of crosslinking.

Function in Food
These modified starches are thickeners that provide short, salve-like textures in processed foods. The choice of starch will depend on processing conditions – heat, acid and shear – in order to achieve adequate granule swelling for optimal viscosity development. These starches produce pastes with fast meltaway.

Benefits
Crosslinked starches offer heat, acid and shear stability. They provide stable viscosity in a wide range of heat processes where native starches would break down with a significant loss of viscosity. Consequently, they can be used at lower dosage rates than their parent native starches.

Limitations
Distarch phosphates are not recommended in chilled or frozen applications. Such products would require both crosslinked and stabilised starches, such as acetylated distarch adipates or hydroxypropylated distarch phosphates.

Within the EU, distarch phosphates are included in Group 1 of Part C in Regulation 1129/2011, additives permitted at *quantum satis*, except in foods for young children where they are permitted to a maximum of 50 000 mg/kg.

Typical Products
Bottled sauces, salad dressings, dry mix puddings and baked goods.

Number	Product
E 1413	Phosphated distarch phosphate

Sources
Native starches are modified by a combination of treatments, as for mono-starch phosphates and distarch phosphates, on an aqueous slurry to produce phosphated distarch phosphates. Typical reagents include orthophosphoric acid, sodium or potassium orthophosphate, sodium tripolyphosphate with phosphorus oxychloride or sodium trimetaphosphate.

Function in Food
These modified starches are used as freeze–thaw stable thickeners. The choice of starch will depend on processing conditions – heat, acid and shear – in order to achieve adequate granule swelling for optimal viscosity and development. These starches produce pastes with fast meltaway.

Benefits
Crosslinked and stabilised starches improve the product quality and shelflife stability of foods. These modified starches offer greater process stability and low-temperature storage stability than their parent native starches and consequently can be used at a lower dosage rate.

Limitations
Within the EU, phosphated distarch phosphates are included in Group 1 of Part C of Regulation 1129/2011, additives permitted at *quantum satis*, except in foods for young children where they are permitted to a maximum of 50 000 mg/kg.

Typical Products
Bottled sauces, pie filling and frozen gravies.

Number	Product
E 1414	Acetylated distarch phosphate

Sources
Native starches are crosslinked and stabilised by reacting an aqueous slurry with reagents such as phosphorus oxychloride or sodium trimetaphosphate

combined with acetic anhydride or vinyl acetate. A range of modified starches is available with different levels of crosslinking and stabilisation.

Function in Food
These modified starches are thickeners and stabilisers, which are designed to maintain granular integrity throughout processing to provide short, salve-like textures in processed foods. The textural qualities are retained even after the processed foods have been chilled or frozen. The choice of starch will depend on processing conditions – heat, acid and shear – in order to achieve adequate granule swelling for optimal viscosity development. These starches produce pastes with fast meltaway.

Benefits
Acetylated distarch phosphates provide stable viscosity in a wide range of heat processes where native starches would break down with a significant loss of viscosity. They also increase shelflife by providing low-temperature stability for chilled and frozen foods. In such applications, native starches, particularly amylase-containing starches would limit shelflife, owing to retrogradation or syneresis.

Limitations
Within the EU acetylated distarch phosphates are included in Group1 Part C of Regulation 1129/2011, additives permitted at *quantum satis*, except in foods for young children where they are permitted to a maximum of 50 000 mg/kg.

Typical Products
Soups, sauces, fruit fillings, chilled and frozen recipe dishes and pet foods.

Number	Product
E 1420	Acetylated starch

Sources
Acetylated starches are produced from native starches by reacting an aqueous slurry with acetic anhydride or vinyl acetate.

Function in Food
Starch acetates, when cooked in water, rapidly develop a stable viscosity with a reduced tendency to set back or retrograde on cooling. Paste clarity is also improved.

Benefits
Acetylated starches are easier to cook, owing to lowering of the gelatinisation temperature. This is a particular benefit for high-amylose starches, which are difficult to cook at ambient pressure, whereas acetylation renders them dispersible under such conditions. The high-solids environment of

confectionery products limits the scope for viscosity control unless an easy-cook thickener, such as acetylated starch, is used. Shelflife stability, particularly in chilled and frozen products, is extended with starch acetates.

Limitations
Acetylated starches often cause curdling in dairy products. This has been attributed to the instability of the acetate linkage in high protein concentrations. In applications such as these, hydroxypropylated starches are more compatible with milk proteins and are therefore recommended for greater stability. Stabilised and crosslinked starches, such as acetylated distarch phosphates or adipates, are more suitable for a wider range of heat-processed foods than starch acetates. Acetylated starches are not as freeze–thaw stable as hydroxypropylated starches.
In the EU acetylated starch is included in Group 1 Part C of Regulation 1129/2011, additives permitted at *quantum satis*, except in foods for young children where they are permitted to a maximum of 50 000 mg/kg.

Typical Products
Batters and breadings, snacks and confectionery products.

Number	Product
E 1422	Acetylated distarch adipate

Sources
Native starches are esterified with acetic anhydride and adipic anhydride to produce acetylated distarch adipates. A range of starches is available with different levels of crosslinking and stabilisation.

Function in Food
The crosslinked and stabilised starches are the most commonly used starch-based thickeners and stabilisers in processed foods. They are designed to provide viscosity stability and tolerance to heat, acid, shear and low-temperature storage. These starches produce pastes with fast meltaway.

Benefits
These modified starches allow the manufacture and distribution of high-quality processed foods. Cooked starch pastes have a higher cold viscosity, owing to their ability to maintain starch structure after processing. Heat penetration is enhanced with more highly crosslinked starches and shelflife is extended, particularly in chilled and frozen products.

Limitations
Acetylated starches often cause curdling in dairy products. This has been attributed to the instability of the acetate linkage in high protein

concentrations. In applications such as these, hydroxypropylated starches are more compatible with milk proteins and are therefore recommended for greater stability. Acetylated starches are not as freeze–thaw stable as hydroxypropylated starches.

Within the EU acetylated distarch adipate is included in Group 1 of Part C of Regulation 1129/2011, additives generally permitted at *quantum satis*, except in foods for young children where they are permitted to a maximum of 50 000 mg/kg.

Typical Products
Gravies, soups, sauces, fruit preparations, sweet and savoury fillings.

Number	Product
E 1440	Hydroxypropyl starch

Sources
Hydroxypropyl starches are made by reacting native starches with propylene oxide. They represent an alternative range of stabilised starches.

Function in Food
These modified starches bind water at lower temperatures than their parent native starches to texturise certain foods and introduce low-temperature stability. Hydroxypropyl starches are frequently further modified to increase their range of applications. The most common combination treatment is crosslinking, where crosslinked and stabilised starches are ideal in chilled and frozen foods.

Benefits
Hydroxypropylated starches are easier to cook than their parent native starches, owing to the reduction in gelatinisation temperature. This is ideal in low-moisture products or high-solids cooking, where competition for water makes it difficult to cook the starch fully and therefore develop maximum stable viscosity.

Limitations
Within the EU, hydroxypropyl starch is included in Group 1 of Part C of Regulation 1129/2011, additives permitted at *quantum satis*.
It is not permitted in foods for babies and young children.

Typical Products
Meats, beverages and low-fat and low-calorie products.

Number	Product
E 1442	Hydroxypropyl distarch phosphate

Sources
These modified starches are produced from the reaction of an aqueous starch suspension with a combination of propylene oxide and either sodium

trimetaphosphate or phosphorus oxychloride. The latter two reagents are crosslinking agents, whilst the former stabilises the starch by etherification. A range of starches is available with different levels of crosslinking and stabilisation.

Function in Food
These crosslinked and stabilised starches are widely used as thickeners, stabilisers and mouthfeel enhancers. The choice of starch will depend on processing conditions – heat, acid and shear – in order to achieve adequate granule swelling for optimal viscosity development and stability. The modified starches also confer excellent low-temperature and freeze–thaw stability. They produce thick, rich, creamy pastes with excellent mouthfeel and cling.

Benefits
The mouthfeel of these starches improves the product aesthetics of low-fat/low-calorie or fat-free products. Their high viscosity allows lower usage rates than is the case with native starches, and they can give a number of processing benefits results from rapid cooking and low fouling of process plant.

Limitations
Hydroxypropyl distarch phosphates are permitted in Group 1 Part C of Regulation 1129/2011, additives permitted at *quantum satis*. These starches are not permitted in foods for babies or young children.

Typical Products
Gravies, soups, sauces, mayonnaises and dressings, sweet and savoury fillings, fruit preparations, dairy products, chilled and frozen recipe dishes and pet foods.

Number	Product
E 1450	Starch sodium octenyl succinate

Sources
Starch sodium octenyl succinate is made by reaction of native starches with 1-octenylsuccinic anhydride (OSA).

Function in Food
These modified starches are effective emulsion stabilisers, owing to the introduction of a hydrophobic moiety onto the starch polymer.

Benefits
Low-viscosity OSA-treated starches can be used at higher solids levels in spray-dried applications, which improves plant efficiency by reducing drying times. These modified starches have better film-forming properties, which results in better oxidation stability and therefore shelflife stability. Emulsified products

benefit from these properties, which improve products aesthetics and extend shelflife. OSA-treated starches are also designed for low-temperature storage, giving temperature tolerance and flexibility in handling flavour emulsions.

Limitations
Starch sodium octenyl succinate is permitted in the EU in Regulation 1129/2011 Part C Group 1, additives permitted at *quantum satis*. It has been allocated an ADI of not specified by JECFA.

Typical Products
Spray-dried flavours, beverage emulsions, emulsified sauces and mayonnaises.

Number	Product
E 1451	Acetylated oxidised starch

Sources
Acetylated oxidised starches are made by modifying native starches with oxidising agents, such as hypochlorite, followed by acetylating agents, such as acetic anhydride.

Function in Food
The modified starches are used as binding and gelling agents in confectionery. Oxidised starches have lower gelatinisation temperatures and hot viscosity, and improved paste clarity and low-temperature-storage stability. The combined acetylation treatment enhances these properties.

Benefits
The starches can be used as alternatives to gelatine and gum arabic in confectionery products.

Limitations
Acetylated oxidised starch is permitted in Group 1 Part C of Regulation 1129/2011 in the EU, additives permitted at *quantum satis*.

Typical Products
Soft sugar confectionery.

Number	Product
E 1452	Starch aluminium octenyl succinate

Sources
Starch aluminium octenyl succinate is made by reaction of aluminium salts with native starches and 1-octenylsuccinic anhydride.

Function in Food

Starch aluminium octenyl succinate is used as an anticaking agent with its major use in cosmetics.

Benefits

Unusually for a modified starch, this compound is hydrophobic and this property is used to stop microencapsulated vitamins and carotenoids sticking together when being dried at low temperature.

Limitations

In the EU, starch aluminium octenyl succinate is permitted in Regulation 1130/2011 in food supplements as defined in Directive 2002/46/EC due to its use in vitamin preparations for encapsulation purposes only, to a maximum of 35 000 mg/kg in the final product.

It has been allocated an ADI by JECFA of not specified.

Number	Product
E 1505	Triethyl citrate

Sources

Triethyl citrate is made by reacting citric acid with ethanol. It is an odourless and colourless oily liquid.

Function in Food

Triethyl citrate is used to increase the rate at which rehydrated egg white powder forms a stable foam. It can also be used as an antifoaming agent, sequestrant, stabiliser and as a carrier solvent.

Limitations

In the EU, triethyl citrate is limited in Regulation 1129/2011 at *quantum satis* in dried egg white, processed eggs and egg products, and to a maximum of 3500 mg/kg in food supplements in capsule and tablet form.

Number	Product
E 1517	Glyceryl diacetate
	Diacetin

Sources

Glyceryl diacetate consists predominantly of a mixture of the 1,2- and 1,3-diacetates of glycerol, with minor amounts of the mono- and triester. It is soluble in water and miscible with ethanol.

Function in Food

Glyceryl diacetate is used as a carrier solvent for flavourings.

Limitations
It is permitted in the EU as a carrier in flavourings in Regulation 1130/2011 to a maximum of 3000 mg/kg.

Number	Product
E 1518	Glyceryl triacetate
	Triacetin

Sources
Glyceryl triactetate is also known as triacetin. It is a colourless oily liquid prepared by reaction of glycerol with acetic anhydride. It is sparingly soluble in water and soluble in ethanol.

Function in Food
Glyceryl triacetate is a hydrophobic liquid used as a lubricant in chewing gum.

Limitations
Glyceryl triacetate is only slightly soluble in water. Within the EU, glyceryl triacetate is permitted in Regulation 1129/2011 at *quantum satis* in chewing gum and in Regulation 1130/2011 as a carrier for additives and flavourings to a maximum of 3000 mg/kg from all sources in foodstuffs individually or in combination with E 1505 and E 1517.

Number	Product
E 1519	Benzyl alcohol

Sources
Benzyl alcohol is a natural constituent of a number of plants, for example, in some edible fruits and in green and black tea. It is approved as a flavouring substance and is used in pharmaceuticals and cosmetics. The material of commerce is synthetic.

Function in Food
Benzyl alcohol is used as a carrier solvent for flavourings.

Limitations
Within the EU, benzyl alcohol is permitted in Regulation 1130/2011 as a carrier in flavourings used in liqueurs and wine products to a maximum of 100 mg/l and in flavourings used in confectionery and bakery wares to a maximum of 250 mg/kg.
Benzyl alcohol has been included in the group ADI for all benzoates of 5 mg/kg by JECFA.

Number	Product
E 1520	Propane-1,2-diol
	Propylene glycol

Sources
Propane-1,2-diol is produced by heating glycerol with sodium hydroxide, or by reacting propylene with chlorinated water to form chlorhydrin followed by treatment with sodium carbonate solution to form the glycol.

Function in Food
Propane-1,2-diol is used as a carrier for enzymes and flavourings.

Limitations
Within in the EU, propane-1,2-diol is permitted in Regulation 1130/2011 as a carrier for enzymes and flavourings up to a maximum of 500 g/kg in enzymes and a total of 3000 mg/kg individually or in combination with E 1505, E 1517 and E 1518.

Number	Product
E 1521	Polyethylene glycol

Sources
Polyethylene glycols (PEG) are made by polymerisation of ethylene oxide and water. The specification for polyethylene glycol covers 6 compounds: PEG 400, 3000, 3350, 4000, 6000 and 8000. The numbers refer to the molecular weight of the compound. These glycols are white waxy substances, soluble in water and practically insoluble in alcohol.

Function in Food
The PEGs are used as glazing agents in film-coating formulations in food-supplement tablets and capsules.

Limitations
Within the EU, polyethylene glycol is permitted in Regulation 1129/2011 at *quantum satis* in table-top sweeteners in tablet and powder form and to a maximum of 10 000 mg/kg in food supplements in capsule and tablet form.

Subject Index

Illustrations and figures are in **bold**. Tables are in *italics*.